# 6th European Academic Symposium on EAF Steelmaking 2025 (EASES 2025)

Journal of Physics: Conference Series Volume 3050

Ljubljana, Slovenia
11 - 13 June 2025

ISBN: 979-8-3313-2384-4
ISSN: 1742-6588

**Printed from e-media with permission by:**

Curran Associates, Inc.
57 Morehouse Lane
Red Hook, NY 12571

CURRAN ASSOCIATES INC. proceedings.com

**Some format issues inherent in the e-media version may also appear in this print version.**



Institute of Physics
Dirac House, Temple Back
Bristol BS1 6BE UK

Phone: 44 1 17 929 7481
Fax: 44 1 17 920 0979

techtracking@iop.org

**Additional copies of this publication are available from:**

Curran Associates, Inc.
57 Morehouse Lane
Red Hook, NY 12571 USA
Phone: 845-758-0400
Fax: 845-758-2633
Email: curran@proceedings.com
Web: www.proceedings.com

# TABLE OF CONTENTS

# Preface to the Proceedings of the 6th European Academic Symposium on EAF Steelmaking

**Vito Logar[1*], Thomas Echterhof[2]**

[1] Laboratory of Control Systems and Cybernetics, Faculty of electrical engineering, University of Ljubljana, Tržaška 25, SI-1000 Ljubljana, Slovenia
[2] Department for Industrial Furnaces and Heat Engineering, RWTH Aachen University, Kopernikusstraße 10, 52074 Aachen, Germany

*E-mail: vito.logar@fe.uni-lj.si

**Abstract.** The European Academic Symposium on EAF Steelmaking (EASES) is a biennial event dedicated to advancing research and development in electric arc furnace (EAF) steelmaking. The 6th edition of this symposium was held from June 11th to 13th, 2025, at the University of Ljubljana, Slovenia. Organized by the Laboratory of Control Systems and Cybernetics at the University of Ljubljana and the Department for Industrial Furnaces and Heat Engineering at RWTH Aachen University, EASES 2025 aimed to provide a platform for researchers, doctoral candidates as well as industry professionals to present and discuss their current topics, research, or projects in the field of EAF steelmaking. A total of 27 presentations were given at the symposium, out of which 12 were accompanied also by a full paper published in IOP Science Journal of Physics: Conference Series (JPCS). The symposium welcomed around 50 participants from academia and industry. The topics that were covered included areas such as modelling, simulation and efficiency of the EAFs, introduction of fossil-free materials, $CO_2$ reduction strategies as well as novel approaches to EAF control.

## 1. Introduction

The European Academic Symposium on EAF Steelmaking (EASES) was initiated at the Department for Industrial Furnaces and Heat Engineering, RWTH Aachen University with the aim of bringing together researchers focused on electric arc furnace (EAF) steelmaking. Following four successful editions held in 2015, 2016, 2018, and 2021 in Aachen, Germany, EASES has grown into a recognized platform for academic and industry experts to exchange insights, share research developments, and foster collaboration in this evolving field. In 2023, the symposium marked a significant milestone by being organized for the first time outside of Aachen, with the 5th EASES held at the University of Oulu in Finland.

Following the established path, the 6th EASES 2025 was the latest instalment in this series of events. It was held in Ljubljana, Slovenia, from June 11th to 13th, 2025. The symposium continued to serve as a dedicated platform for scientific exchange and collaboration in the field of EAF steelmaking, with particular emphasis on supporting early-stage researchers and doctoral students.

The conference was jointly organized by the Laboratory of Control Systems and Cybernetics at the University of Ljubljana and the Department for Industrial Furnaces and Heat Engineering at RWTH Aachen University. The event was co-chaired by Vito Logar (Faculty of Electrical

Engineering, University of Ljubljana) together with Thomas Echterhof (Department for Industrial Furnaces and Heat Engineering, RWTH Aachen University). The conference was made possible thanks to the collaboration and support of the local organizing committee at the Faculty of Electrical Engineering, who played a vital role in its organization (Prof. Igor Škrjanc, Asst. Prof. Simon Tomažič, Dr. Goran Andonovski, Dr. Aljaž Blažič, and Assoc. Prof. Vito Logar). The success of EASES 2025 is underpinned also by the guidance and expertise of a dedicated Scientific Committee, comprised of leading academics and industry experts in EAF steelmaking from Finland, Germany, Italy, and Slovenia. The committee played a vital role in shaping the conference program, ensuring scientific integrity, and maintaining the symposium's reputation as a notable scientific event in this field (Table 1).

**Table 1.** Scientific committee of EASES 2025.

| Member | Affiliation | Country |
|---|---|---|
| Dr. Matti Aula | Outokumpu Steel Oy | Finland |
| Dr.-Ing. Thomas Echterhof | RWTH Aachen University | Germany |
| Assoc. Prof. Vito Logar | University of Ljubljana | Slovenia |
| Assoc. Prof. Davide Mombelli | Politecnico di Milano | Italy |
| Assoc. Prof. Ville-Valtteri Visuri | University of Oulu | Finland |

## 2. Contributions and Review Process

EASES 2025 symposium showcased 27 technical presentations, highlighting the latest developments and research in topics ranging from energy optimization and process control to sustainable practices and material innovations. Of these, 12 full papers have been peer-reviewed and accepted for publication in the official proceedings.

The Call for Papers for the conference was announced in October 2024. Initially, the deadline for submissions was set for January 17$^{th}$, 2025, and later extended until January 31$^{st}$, 2025. The deadline for submitting full papers was set for April 14$^{th}$, 2025. The review process was completed by May 26$^{th}$, 2025.

The review process for the proceedings was single-blind and adhered to rigorous academic standards, resulting in a 12 % rejection rate. We extend our sincere appreciation to all reviewers for their timely and constructive feedback, which ensured the high quality and relevance of the published contributions.

## 3. Thematic Scope of EASES conference series

The thematic scope of the EASES conference series reflects the rapidly evolving challenges and innovations in EAF steelmaking, covering:

- **Efficiency and Process Optimization**
  Research in this area targets improved energy usage, shorter tap-to-tap times, and more consistent process control, aiming to make EAF operations more economically and environmentally sustainable.

Journal of Physics: Conference Series **3050** (2025) 011001 doi:10.1088/1742-6596/3050/1/011001

- **Process Control and Sensors**
  Developments in digital monitoring, sensor integration, and real-time control systems are enhancing precision, safety, and responsiveness within the furnace environment.
- **$CO_2$ Emissions Reduction and Environmental Impact**
  The steel industry is under increasing pressure to decarbonize, and EAF-specific solutions—including greener electricity use and optimized material input—are at the forefront of current efforts.
- **Process modelling and simulation**
  Computational modelling enables researchers to analyse complex thermal and chemical behaviours, predict process outcomes, and support data-driven decision-making.
- **Scrap Treatment, Alternative/new charge materials and Control**
  Innovations in scrap sorting, pre-treatment, and novel charge materials such as DRI and HBI are helping to improve material efficiency and product quality.
- **Development of New Steel Grades**
  With growing demands for specialized steels, research is advancing the metallurgical design and process routes for high-performance and tailored steel products.
- **Slag Valorisation and By-product Management**
  Strategies for recycling and repurposing slag and other by-products are contributing to waste reduction, resource efficiency, and circular economy practices in EAF operations.

## 4. Technical presentations

The EASES 2025 symposium began on June 12th, 2025, with a welcome speech by the symposium co-chair Vito Logar (University of Ljubljana). Afterwards, technical sessions began with fifteen presentations on the first day (Table 2).

**Table 2:** Technical presentations on June 12th, 2025

| | |
|---|---|
| **Session on EAF modelling, simulation and efficiency 1** | |
| 1 | **Sensitivity Analysis of State Space Models for Scrap Composition Estimation in EAF and BOF**<br>*Yiqing Zhou, Karsten Naert, Dirk Nuyens* |
| 2 | **Application of an artificial neural network to estimate the off-gas generation within an EAF under modified operating conditions**<br>*Alexander Reinicke, Lilly Schulte, Thomas Echterhof* |
| 3 | **An Integrated CFD Modeling Approach Towards an Entire EAF Operation Process**<br>*Chenn Zhou, Orlando Ugarte, Tyamo Okosun, Shiyu Wang, Sathvika Kottapalli, Joe Maiolo, Hamzah Alshawarghi* |
| 4 | **Foam-Arc Interaction in Electric arc Furnace: Insights into Flow, Stability, and Thermal Behaviour**<br>*Mohamad Al Nasser, Ebrahim Karimi Sibaki, Menghuai Wu, Anton Ishmurzin, Gernot Hackl, Nikolaus Voller, Christian Redl, Harald Holzgruber, Kharicha Abdellah* |

Conference program continued on the June 13th, 2025, with a welcome speech by the symposium co-chair Thomas Echterhof (RWTH Aachen University) and the remaining sessions (Table 3).

**Table 3:** Technical presentations on June 13th, 2025

**Session on Fossil-free raw materials**

16 **Kinetic Modeling of Hematite Reduction by Hydrogen Plasma Smelting Reduction in laboratory scale**
*Areej Javed, Ilpo Mäkelä, Henri Pauna, Henna-Riikka Putaala, Ubaid Manzoor, Dennis Klapproth, Isnaldi R. Souza Filho, Ville-Valtteri Visuri*

17 **Mass and energy based modelling of EAF steelmaking scenarios using scrap and hydrogen reduced DRI as raw materials**
*Eetu-Pekka Heikkinen, Petri Sulasalmi, Ville-Valtteri Visuri, Seppo Ollila, Jarmo Lilja*

18 **Optimization Based Experimental Design of Metal-Slag Experiments in Hydrogen Plasma Smelting Reduction Process**
*Tero Vuolio, Ville-Valtteri Visuri, Michael Zarl, Iivari Lappeteläinen*

19 **Modeling the thermal properties of H-DRI during melting in an electric arc furnace**
*Ankur Agnihotri, Petri Sulasalmi, Ville-Valtteri Visuri*

**Session on Process control and sensors**

20 **Implementation of an At-Line Slag Analyzer – Advantages and Challenges**
*Alexander Schlemminger*

21 **Model Predictive Control of the Refining Stage in an Electric Arc Furnace Using Soft Sensors**
*Aljaž Blažič, Igor Škrjanc, Vito Logar*

**Session on CO2 emission reduction and environmental impact 1**

22 **Scope 3 emissions in secondary steelmaking: relevance and impact on CFP**
*Luca Testini, Alessandro Misul, Vincenzo Morreale, Philippe Brocard, Livia Persico, Davide Mombelli, G Dotelli*

23 **Scope 3 emissions in secondary steelmaking: relevance and impact on organization carbon footprint**
*Luca Testini, Alessandro Misul, Vincenzo Morreale, Philippe Brocard, Livia Persico, Davide Mombelli, G Dotelli*

24 **Using a Novel Scaled Injector to Evaluate Biocarbon for Slag Foaming in EAF Steelmaking**
*Christopher DiGiovanni, Tiago Fernandes Lins, Michael Strelbisky, Majid Zamani, Allan Runstedtler, Colin Scott*

**Session on CO2 emission reduction and environmental impact 2**

25 **Decarbonization and New Energy-Efficient Technologies for EAF Steelmaking**
*Hamzah Alshawarghi, Joachim von Schéele*

26 **Results from the Experimental Campaigns with the H2 Oxyfuel Burner for Electric Arc Furnaces**
*Eros Luciano Faraci, Irene Luzzo, Jacopo Greguoldo, Fabio Vecchiet, Giulio Rinaldi, Fabiano Ferrari, Federico Nastro, Daniele Gaspardo, Lilly Schulte*

27 **Numerical Investigation of Hydrogen Blending on the Impinging Flame Structure in Non-Premixed CH4/H2/Air combustion for Scrap Metal Heating**
*Gopal Pandey, Geoffrey Brooks, Jamal Naser, Daniel Liang*

## 5. Proceedings

All authors who submitted abstracts for the symposium and were accepted were also invited to submit a full paper describing a more detailed overview of their work. Each submitted paper was subjected to single anonymous peer-review. After peer-review and revisions, 12 manuscripts out of 17 were accepted to be published in the IOP Science Journal of Physics: Conference Series (JPCS) (Table 4). The proceedings are indexed in all the main indexing services, including Web of Science and Scopus.

**Table 4:** Peer-reviewed manuscripts published as part of the proceedings.

1 **Modeling the thermal properties of H-DRI during melting in an electric arc furnace**
*Ankur Agnihotri, Petri Sulasalmi, Ville-Valtteri Visuri*

2 **Kinetic Modeling of Hematite Reduction by Hydrogen Plasma Smelting Reduction in laboratory scale**
*Areej Javed, Ilpo Mäkelä, Henri Pauna, Henna-Riikka Putaala, Ubaid Manzoor, Dennis Klapproth, Isnaldi R. Souza Filho, Ville-Valtteri Visuri*

3 **Sensitivity Analysis of State Space Models for Scrap Composition Estimation in EAF and BOF**
*Yiqing Zhou, Karsten Naert, Dirk Nuyens*

4 **Application of an artificial neural network to estimate the off-gas generation within an EAF under modified operating conditions**
*Alexander Reinicke, Lilly Schulte, Thomas Echterhof*

5 **Iron Recovery from Waelz Slag through Biogenic Carbothermic Reduction**
*Gianluca Dall'Osto, Davide Mombelli, Sara Scolari, Carlo Mapelli*

6 **Quantitative phase analysis in carbon steel EAF slag for the determination of phase-controlled leaching mechanism**
*Davide Mombelli, Sara Scolari, Gianluca Dall'Osto, Carlo Mapelli*

7 **Using a Novel Scaled Injector to Evaluate Biocarbon for Slag Foaming in EAF Steelmaking**
*Christopher DiGiovanni, Tiago Fernandes Lins, Michael Strelbisky, Majid Zamani, Allan Runstedtler, Colin Scott*

8 **Mass and energy based modelling of EAF steelmaking scenarios using scrap and hydrogen reduced DRI as raw materials**
*Eetu-Pekka Heikkinen, Petri Sulasalmi, Ville-Valtteri Visuri, Seppo Ollila, Jarmo Lilja*

Journal of Physics: Conference Series **3050** (2025) 011001 doi:10.1088/1742-6596/3050/1/011001

9 **Numerical Investigation of Hydrogen Blending on the Impinging Flame Structure in Non-Premixed CH4/H2/Air combustion for Scrap Metal Heating**
*Gopal Pandey, Geoffrey Brooks, Jamal Naser, Daniel Liang*

10 **Smelting of various steel-plant dusts to evaluate recovery of zinc and iron via the Enviroplas process**
*Sello Tsebe, Sanda Moloane, Habib Zughbi, Deside Chibwe, Peter Austin, Dursman Mchabe, Mukhethwa Netshia, Derek Hayman, Elias Matinde*

11 **Optimization Based Experimental Design of Metal-Slag Experiments in Hydrogen Plasma Smelting Reduction Process**
*Tero Vuolio, Ville-Valtteri Visuri, Michael Zarl, Iivari Lappeteläinen*

12 **Model Predictive Control of the Refining Stage in an Electric Arc Furnace Using Soft Sensors**
*Aljaž Blažič, Igor Škrjanc, Vito Logar*

## 6. Concluding remarks

The Electric Arc Furnace process continues to be a major research focus in the field of ferrous metallurgy. Following the successful events held in Germany and Finland, the 6th EASES in Ljubljana, Slovenia, continued the envisioned path in the internationalization and expansion of the symposium.

The symposium included 27 presentations accompanied by 12 full papers. Before the event, the authors who decided to publish a full paper were invited to submit their manuscripts for inclusion in the official proceedings. Following a peer-review process, 12 out of 17 papers were ultimately accepted for publication in the conference proceedings.

We hope that the insights shared, and connections formed at EASES 2025 will contribute meaningfully to ongoing research and collaboration in the field of EAF steelmaking, and we look forward to seeing how this collective work will shape future advancements.

## Acknowledgement

The authors would like to thank the members of the scientific committee for their invaluable contribution to the EASES organization as well as to the local organizing committee making EASES 2025 another successful event.

Journal of Physics: Conference Series **3050** (2025) 011002 doi:10.1088/1742-6596/3050/1/011002

# Peer Review Statement

All papers published in this volume have been reviewed through processes administered by the Editors. Reviews were conducted by expert referees to the professional and scientific standards expected of a proceedings journal published by IOP Publishing.

- **Type of peer review:** Single Anonymous
- **Conference submission management system:** Morressier
- **Number of submissions received:** 17
- **Number of submissions sent for review:** 17
- **Number of submissions accepted:** 15
- **Acceptance Rate (Submissions Accepted / Submissions Received × 100):** 88.2
- **Average number of reviews per paper:** 1
- **Total number of reviewers involved:** 5
- **Contact person for queries:**
  **Name:** Vito Logar
  **Email:** vito.logar@fe.uni-lj.si
  **Affiliation:** University of Ljubljana, Faculty of electrical engineering

Journal of Physics: Conference Series **3050** (2025) 012001 doi:10.1088/1742-6596/3050/1/012001

# Sensitivity Analysis of State Space Models for Scrap Composition Estimation in EAF and BOF

**Yiqing Zhou[1], Karsten Naert[2] and Dirk Nuyens[1]**

[1]Department of Computer Science, KU Leuven, Leuven, Belgium
[2]ArcelorMittal, Gent, Belgium

E-mail: yiqing.zhou@kuleuven.be, karsten.naert@arcelormittal.com, dirk.nuyens@kuleuven.be

**Abstract.** This study develops and analyzes linear and nonlinear state space models for estimating the elemental composition of scrap steel used in steelmaking, with applications to Electric Arc Furnace (EAF) and Basic Oxygen Furnace (BOF) processes. The models incorporate mass balance equations and are fitted using a modified Kalman filter for linear cases and the Unscented Kalman Filter (UKF) for nonlinear cases. Using Cu and Cr as representative elements, we assess the sensitivity of model predictions to measurement noise in key process variables, including steel mass, steel composition, scrap input mass, slag mass, and iron oxide fraction in slag. Results show that the models are robust to moderate noise levels in most variables, particularly when errors are below 10%. However, accuracy significantly deteriorates with noise in slag mass estimation. These findings highlight the practical feasibility and limitations of applying state space models for real-time scrap composition estimation in industrial settings.

## 1 Introduction

Steel is a foundational material in modern society, with applications ranging from construction and transportation to household appliances. However, steel production is a significant source of greenhouse gas emissions, presenting a major environmental challenge. As global demand for steel continues to rise, reducing the industry's environmental footprint has become increasingly urgent. One promising approach is the recycling of scrap steel, which can substantially lower energy consumption, reduce waste, and decrease carbon emissions. Nonetheless, integrating scrap steel into the production process remains complex due to the variable and unpredictable composition of scrap material.

The Basic Oxygen Furnace (BOF) and Electric Arc Furnace (EAF) are two common methods for producing steel, which can make use of recycled scrap. Although, using recycled scrap helps to reduce environmental impact, the unpredictable composition of scrap steel makes this a hard problem. Changes in the scrap's elemental fraction can cause issues with product quality, make the process harder to control, and complicate predictions about the process [3, 6]. If the uncertainty of scrap composition is high, only a limited portion of scrap can be used in the charge, requiring a greater reliance on iron ore and consequently increasing $CO_2$ emissions. Therefore, accurate estimation of the elemental fraction of different scrap types is essential for reducing this uncertainty. By enhancing the accuracy of scrap composition estimation, we can safely increase its usage in steelmaking and further reduce the industry's environmental footprint.

New techniques for modeling and improving these processes are helping to overcome these challenges. Methods such as least squares estimation [11, 2], maximum likelihood estimation [1], and stochastic optimization [5] have been used to predict the amount of different elements in scrap and the uncertainty

Journal of Physics: Conference Series **3050** (2025) 012001 doi:10.1088/1742-6596/3050/1/012001

in those predictions. However, these methods often struggle to deal with changes in scrap composition over time, limiting their use in real-time control.

This study builds upon a novel framework that addresses these limitations through the use of state-space models and advanced filtering techniques. Specifically, Kalman filters and unscented Kalman filters are employed to estimate the evolving elemental composition of scrap steel in real-time, which can be found in [12]. Rather than focusing on the detailed mathematical formulation of the models and algorithms, which have been thoroughly discussed in [12], this work emphasizes the sensitivity of estimation performance to measurement noise across different sensor inputs.

We use copper (Cu) and chromium (Cr) as representative examples to illustrate the linear and non-linear models respectively. Although keeping the Cr fraction within bounds is normally not such a problem for BOF processes, unless for very particular types of steel production, it is more of a problem for EAF processes when limited amounts of clean scrap are available. In any case, Cr is chosen as an example element because it partitions between steel and slag, introducing nonlinearity into the model. The methodology is general and can be applied to other elements that either retain in the steel or partition into the slag.

The remainder of this paper is organized as follows: Section 2 introduces a linear and a non-linear state-space model for two different types of elements, and corresponding filtering algorithms to fit the models. Section 3 analyzes the sensitivity of estimation accuracy to measurement noise in various settings. Section 4 concludes the paper with a summary of findings and potential directions for future work.

## 2 State space models and corresponding filtering algorithms

This section reintroduces the state space model and associated filtering algorithms presented in [12], with the aim of clarifying the modeling framework and the underlying assumptions.

### *2.1 State space models*

In the steelmaking process, certain chemical elements only go into the steel, while others partition between the steel and the slag. The model for these two kinds of elements are different. Various process variables are measurable in practice, such as the input mass of each scrap type, the total mass of tapped steel, and the mass fraction of an element X of interest in the steel. Some other variables, including the mass of slag and the iron oxide content in slag, are often estimated using empirical models.

For elements that remain only in the steel, the observation model can be directly written as a linear function of the state variables, which represent the elemental composition of the scrap types. This results in a linear model. However, for elements that partition between steel and slag, it is often challenging to measure or estimate all the necessary process variables, such as the elemental fraction in slag or the partition coefficients. To address this, the model is extended by introducing additional state variables to estimate the necessary process variables, allowing the observation to be expressed in a form consistent with the mass balance principles. Thus the observation is no longer a linear function of the state variable, which makes it a non-linear model.

*Case 1: Element $X$ retained in steel only.* For an element $X$ that enters only the steel, the system can be modeled using the following linear state space formulation for heats $t = 1, 2, \ldots,$

$$\vec{\alpha}_{t+1} = (1-\gamma)\,\vec{\alpha}_t + \gamma\,\vec{\eta}_t, \qquad \vec{\eta}_t \sim \mathcal{D}(\vec{q}, Q), \tag{1}$$

$$y_t = \vec{m}_t \cdot \vec{\alpha}_t + \varepsilon_t, \qquad \varepsilon \sim \mathcal{N}(0, H_t). \tag{2}$$

The variables in this model are defined as follows.

- $\vec{\alpha}_t$: The fraction of element $X$ across scrap types at heat $t$; this is the unobserved state variable.
- $\gamma$: A small positive constant representing temporal correlation between adjacent heats.
- $\vec{\eta}_t$: A random variable that introduces temporal variability into $\vec{\alpha}_t$ and reflects the long-term average fraction of element $X$ in each scrap type. Each component of $\vec{\eta}_t$ is sampled from a Beta distribution, with its mean given by the corresponding entry in $\vec{q}$ and its variance determined by the corresponding diagonal entry in the diagonal matrix $Q$.
- $\vec{m}_t$: Input mass of each scrap type at heat $t$.
- $y_t$: Observed total mass of element $X$ from scrap at heat $t$. It can be calculate by other measurements using $m_{\text{steel},t} f_{\text{steel,X},t} - m_{\text{hm},t} f_{\text{hm,X},t}$.

- $\varepsilon_t$: Observation noise, modeled as zero-mean Gaussian noise with covariance $H_t$.

Equation (1) governs the state dynamics, while Equation (2) models the observations based on mass balance assumptions.

*Case 2: Element X partitioning between steel and slag.* For elements that are distributed across both steel and slag, the observation model becomes nonlinear due to the partition behavior. The extended state space model is given by

$$\vec{\alpha}_{t+1}^{+} = \begin{bmatrix} \vec{\alpha}_{t+1} \\ \vec{c}_{t+1} \end{bmatrix} = (1-\gamma) \begin{bmatrix} \vec{\alpha}_{t} \\ \vec{c}_{t} \end{bmatrix} + \gamma \begin{bmatrix} \vec{\eta}_{t} \\ \vec{\theta}_{t} \end{bmatrix}, \qquad \vec{\eta}_t \sim \mathcal{D}(\vec{q}, Q), \quad \vec{\theta}_t \sim \mathcal{D}(\vec{q}_c, Q_c), \tag{3}$$

$$y_t = \frac{\vec{m}_t \cdot \vec{\alpha}_t + m_{\mathrm{hm},t} f_{\mathrm{hm},t}}{1 + (c_{1,t} + c_{2,t} f_{\mathrm{FeOn,\ slag},t})\, m_{\mathrm{slag},t}/m_{\mathrm{steel},t}} + \varepsilon_t, \qquad \varepsilon \sim \mathcal{N}(0, H_t). \tag{4}$$

The new variables are listed below.

- $\vec{\alpha}_t^{+}$: The extended state vector, with $\vec{c}_t$ representing the partition coefficients for element $X$.
- $\vec{\theta}_t$: The corresponding random variable for $\vec{c}_t$, where each component follows a Beta distribution with mean and variance be the corresponding component of vector $\vec{q}_c$ and of diagonal matrix $Q_c$.
- $y_t$: Observed mass of element $X$ in the steel, computed via a nonlinear mass balance formula.
- Other terms: mass of hot metal $m_{\mathrm{hm},t}$, fraction of X in hot metal $f_{\mathrm{hm},t}$, fraction of iron oxide in slag $f_{\mathrm{FeOn,slag},t}$, the mass of slag $m_{\mathrm{slag},t}$, and the mass of steel $m_{\mathrm{steel},t}$.

The state update equation (3) is similar to (1) for the linear model, while the observation equation (4) is no longer a linear combination of state variable $\vec{\alpha}_t^{+}$ anymore, which makes it a non-linear model.

### *2.2 Filtering algorithms*

This section introduces filtering algorithms suitable for the linear and nonlinear state space models described above. The Kalman filter is used for the linear case, while the Unscented Kalman Filter (UKF) is adopted for the nonlinear model.

*Linear Model: Kalman Filter* For the linear state space model (1)–(2), the Kalman filter provides a standard approach to estimate the state from noisy observations. It combines model predictions with measurements to yield updated state estimates over time. Although $\vec{\eta}_t$ follows a distribution $\mathcal{D}(\vec{q}, Q)$ constrained within $[0, 1]$, its small variance justifies approximating it as Gaussian. This leads to a slight modification of the standard Kalman filter, which typically assumes zero-mean noise [9, Section 4.3]. In our case, the noise has nonzero mean $\vec{q}$, altering the prediction step accordingly. The parameters and initial state can be estimated using the Expectation-Maximization (EM) algorithm [10], or some other numerical methods. However, since there are other measurements and professional knowledge available for the steelmaking process, it is recommended to manually tune the parameters.

*Nonlinear Model: Unscented Kalman Filter* For the nonlinear model (3)–(4), we apply the Unscented Kalman Filter (UKF) [7, 8], following [4, Section 10.3]. The UKF uses the unscented transform to propagate a set of sigma points through nonlinear functions, capturing the posterior distribution more accurately than linear approximations. Our model differs slightly from classical formulations [9, equations (9.36)–(9.37)] by allowing nonzero mean noise $[\vec{q}, \vec{q}_c]^\top$, which requires small adaptations to the UKF algorithm. Hyperparameters are also tuned manually based on process knowledge.

### *2.3 Input uncertainty analysis*

The model and associated filtering algorithms have been introduced and motivated in [12], which also includes a detailed analysis of model misspecification and supporting numerical results. However, [12] assumes that all input variables are measured precisely, accounting only for observation noise in the form of $\varepsilon_t$ in (2) and (4).

In the following section, we extend this analysis by examining the impact of measurement noise in other variables, such as the mass of input scrap and the mass of steel, on the accuracy and robustness of state estimation and model prediction.

Journal of Physics: Conference Series **3050** (2025) 012001 doi:10.1088/1742-6596/3050/1/012001

## 3 Numerical results

Validation of the algorithms, the impact of model misspecification, and results based on real data are discussed in [12]. This section focuses on evaluating the sensitivity of model predictions with respect to various measurements. The linear model (1)–(2) is demonstrated using Cu as a representative element, while Cr is used for the nonlinear model (3)–(4).

The models rely on several key measurements:

- steel mass $m_{\text{steel},t}$,
- element fraction in steel $f_{\text{steel},\text{X},t}$,
- input mass of scrap types $\vec{m}_t$,
- slag mass $m_{\text{slag},t}$,
- iron oxide fraction in slag $f_{\text{FeOn},\text{slag},t}$.

To evaluate the robustness of the models under realistic measurement uncertainty, we introduce synthetic noise into key process variables using a multiplicative Gaussian noise model

$$\tilde{m} = (1+\xi)\, m, \quad \xi \sim \mathcal{N}(0, c),$$

with noise levels $c \in [0\%, 20\%]$. Since noise in $m_{\text{steel},t}$ and $f_{\text{steel},\text{X},t}$ affects only the observation model (2), their effect can be absorbed into the observation noise covariance $H_t$ and is not analyzed further. While we run some of the experiments for very low, unrealistic, noise levels, like 1%, the focus is on evaluating model performance under controlled levels of synthetic noise, where the 1% noise level then holds as an ideal best case scenario. We expect realistic noise levels to be able to be managed to be under 20%. In practice, the key inputs to the proposed models are obtained from the production process. Scrap mass is typically measured using mobile weighing systems, while slag mass is estimated via empirical models based on correlated process variables. The following experiments hence give an idea of what kind of accuracy is tolerable for these measurements.

### *3.1 Synthetic data generation*

The model was tested on actual BOF data from ArcelorMittal in the context of [12]. Our model has also been run on real EAF data from ArcelorMittal in the context of this paper; however, due to confidentiality constraints, those results cannot be published. Therefore, the results presented in this paper are based on synthetic, yet realistic, data.

To enable a controlled and interpretable analysis, we now generate synthetic data that closely mimics realistic operating conditions. This also allows for a more systematic sensitivity analysis, as the true values of all variables are known. To conduct the sensitivity analysis, synthetic data is generated under controlled conditions. We assume the presence of 45 distinct scrap types and $T = 20000$ heats. The following experiments are designed to mimic BOF process data. According to the models (1)–(2) and (3)–(4), the EAF process can be represented as a special case by setting $m_{\text{hm}} = 0$.

*Elemental fraction of scrap types* For both Cu and Cr, we first generate the initial elemental fraction vector $\vec{\alpha}_1 = \vec{q}$. The 45 scrap types are divided into three groups (15 types per group), with their values sampled from the following uniform distributions with unit [ppm].

- Group 1 (types 1–15): $\mathcal{U}(200, 1000)$.
- Group 2 (types 16–30): $\mathcal{U}(1000, 2000)$.
- Group 3 (types 31–45): $\mathcal{U}(2000, 5000)$.

This grouping reflects common practices in real production, where scrap types vary in impurity levels, some containing more Cu and Cr than others. The covariance matrix $Q$ is diagonal, with elements $Q_{i,i} = 5q_i^2$ for $i = 1, \ldots, 45$. The temporal correlation constant is set to $\gamma = \ln(2)/1000$. The sequence $\vec{\alpha}_t$ for $t > 1$ is then generated recursively using the linear state update (1) or the nonlinear model (3), depending on the element. These choices are motivated in [12].

*Partition coefficients of Cr* For the Cr model, the initial and prior mean of the partition coefficients is set to $\vec{c}_1 = \vec{q}_c = [9.7,\ 0.01]^\top$. The covariance matrix $Q_c$ is diagonal: $Q_c$ is set to $[5 \times 9.7^2,\ 0;\ 0,\ 5 \times 0.01^2]$. The values of $\vec{c}_t$ for $t > 1$ are then generated recursively via the state equation (3).

Journal of Physics: Conference Series **3050** (2025) 012001 doi:10.1088/1742-6596/3050/1/012001

*Input mass of scrap types* We generate a matrix for input mass of scrap types, where the $t$-th row of the matrix represents the value of $\vec{m}_t$, $t = 1, \ldots, T$. The matrix satisfies some conditions: all the elements are non-negative, and the summation of each row is around 70 t. This row sum reflects the practical capacity constraints of a BOF, making it a plausible and realistic. The details about generating the matrix will be discussed in Section 3.2.1.

*Other variables* Table 1 lists the variables whose values are sampled from uniform distributions, along with their respective upper and lower bounds.

Table 1: The lower and upper bound of the uniform distribution for variables.

| Variables | Lower | Upper | Unit |
|---|---|---|---|
| $m_{\text{hm},t}$ | 280 | 290 | [t] |
| $f_{\text{hm,Cu},t}$ | 20 | 30 | [ppm] |
| $f_{\text{hm,Cr},t}$ | 50 | 200 | [ppm] |
| $m_{\text{slag},t}$ | 20 | 30 | [ppm] |
| $m_{\text{steel},t}$ | 340 | 350 | [ppm] |
| $f_{\text{FeOn,slag},t}$ | 20 | 30 | [ppm] |

*Elemental fraction of steel* Since all of the other variables have already been generated, the elemental fraction of steel can be calculated using mass balance. And the standard deviation of measurement error for $f_{\text{steel,Cu},t}$ and $f_{\text{steel,Cr},t}$ are 10 and 5 ppm, respectively.

In the following experiments, we evaluate both the prediction error for the Cu and Cr fractions in steel and the estimated elemental fractions of the scrap types. This means that, in an ideal scenario, the prediction error would have zero mean and a standard deviation of 10 ppm for Cu and 5 ppm for Cr, representing the theoretical best-case performance under the assumed model.

*3.2 Input scrap mass*

*3.2.1 Impact of fixed scrap recipes* In industrial steelmaking, it is common for plants to reuse predefined "recipes" of scrap combinations tailored for specific steel grades. These recipes often consist of fixed proportions of several scrap types, resulting in limited variability in the input mass vectors $\vec{m}_t$ over time. Mathematically, this lack of diversity is reflected in the matrix $M$, constructed by stacking all $\vec{m}_t$ vectors across timesteps, which often becomes rank-deficient or exhibits a high condition number.

To investigate the implications of this structural limitation, we conduct numerical experiments by generating matrices with different properties, while keeping the total scrap input mass similar to the typical value of 70 t. There are different ways of constructing the matrix of $\vec{m}_t$ values corresponding to different scenarios that will be used in the experiments. The following ways will be used.

- **Full-rank matrix with controlled condition number:** This matrix is generated using singular value decomposition to achieve a specific condition number. In practice, fixed scrap recipes can result in such a matrix structure.

- **Low-rank matrix:** A low-rank structure is imposed by multiplying two random matrices of sizes $m \times r$ and $r \times n$. Since rank-deficient matrices have an infinite condition number, they represent a case when fixed recipes are used with very little variability.

- **Sparse matrix:** This matrix is constructed to be sparse, with each row containing at least one non-zero element and the overall density controlled by a specified parameter. After initial sparsification, additional mass is selectively added to a few specific scrap types (e.g., columns 1, 23, and 45), simulating the practical preference for well-characterized and frequently used scrap types.

For all of these methods each row is finally normalized to have a total mass of approximately 70 t. This setup is quite common in real applications, as scrap types with lower uncertainty are often favored in production planning.

The results are summarized in Table 2 and visualized in Figure 1. Each row in the table corresponds to a different matrix $M$. The first row of this table corresponds to a case where only one scrap type is used per heat and, although unrealistic, serves as a best possible case to which the other results can be compared. The second row uses a sparse matrix, and the third row uses a full rank matrix with high condition number. The fourth row uses a low rank matrix. The table reports the mean and standard deviation of the prediction error for elemental fractions in steel. The standard deviation of observing $f_{\text{steel,Cu},t}$ and $f_{\text{steel,Cr},t}$ are 10 ppm and 5 ppm, respectively, representing an approximate lower bound on achievable error. The results in the table approach or even surpass this bound in some cases.

However, a closer look at the figures reveals more nuanced behavior. The red bars show the average input mass of the scrap type grouped over 30 heats. In the second row, the scrap type is used frequently but in small quantities. In contrast, in the other two rows, the scrap type is used consistently and in larger quantities. This provides more information for estimating its composition. Thus the estimated elemental fraction tracks the generated data much better.

These observations suggest an important practical insight: When scrap types appear only in fixed ratios or low amounts, the model struggles to distinguish their individual contributions, relying more heavily on priors and causing estimates to revert toward the prior mean. Therefore, in practice, to improve the estimation of a particular scrap type's elemental fraction, it is advisable to use that scrap type more frequently and in larger amounts.

Table 2: Effect of matrix rank and condition number on predicted Cu and Cr fractions in steel (ppm). The standard deviation of measurement error for $f_{\text{steel,Cu},t}$ and $f_{\text{steel,Cr},t}$ are 10 ppm and 5 ppm, respectively, representing the theoretical best-case performance.

| Rank $(M)$ | Cond $(M)$ | $\mathrm{E}[\varepsilon(f_{\text{steel,Cu},t})]$ | $\sigma[\varepsilon(f_{\text{steel,Cu},t})]$ | $\mathrm{E}[\varepsilon(f_{\text{steel,Cr},t})]$ | $\sigma[\varepsilon(f_{\text{steel,Cr},t})]$ |
|---|---|---|---|---|---|
| 45 (Identity-based matrix) | 1 | −0.17 | 13.96 | −0.14 | 5.34 |
| 45 (Sparse matrix) | 25 | −0.33 | 10.79 | −0.07 | 5.39 |
| 45 (Full rank matrix) | 719104 | 0.75 | 7.59 | −0.12 | 3.63 |
| 20 (Low rank matrix) | $\infty$ | 1.50 | 7.73 | −0.28 | 3.38 |

Since the sparse matrix almost reaches the theoretical best-case performance in Table 2 and also tracks the generated true elemental fraction well in Figure 1, the following experiments are all implemented with the same matrix.

*3.2.2 Sensitivity to measurement noise* Accurate measurement of the input scrap mass $\vec{m}_t$ is essential for estimating the elemental fraction of the steel and scrap. However, in practice, errors may occur due to handling losses, inaccurate weighing, or manual loading deviations. These uncertainties can propagate through the model and affect the quality of the composition estimates. To evaluate the model's robustness, we introduce multiplicative Gaussian noise to $\vec{m}_t$ and assess its effect on the predicted elemental fractions in the steel and scrap types. The results are summarized in Table 3 and visualized in Figure 2.

The findings indicate that, when the measurement error of input scrap mass get larger, the bias and standard deviation of error of predicted elemental fraction in steel also gets larger. The model is relatively robust to small levels of noise. When the noise level remains below 20%, both the bias and standard deviation of the predicted Cu and Cr fractions in steel remain low and stable. For estimating the elemental fraction of scrap, the results are influenced more, compared with the last row of Figure 1. These results demonstrate that the model is robust to realistic levels of measurement noise. Even with noise levels up to 20%, the accuracy for predicting both Cu and Cr fraction in steel remains largely unaffected, with minimal increases in bias or variance, while the estimated Cu and Cr fraction of scrap types are affected.

This suggests that the model can reliably handle the typical uncertainties encountered in industrial scrap handling and weighing, provided that measurement errors are reasonably controlled. Such robustness is particularly valuable for real-time monitoring and decision-making in steelmaking processes, where perfect data quality cannot always be guaranteed.

*3.3 Slag mass*

Slag mass $m_{\text{slag},t}$ plays a critical role in the nonlinear model, particularly for elements such as Cr that partition between steel and slag. In practical settings, however, slag mass is not directly measured but estimated from other process variables or inferred using empirical models, making it inherently uncertain.

Journal of Physics: Conference Series **3050** (2025) 012001 doi:10.1088/1742-6596/3050/1/012001

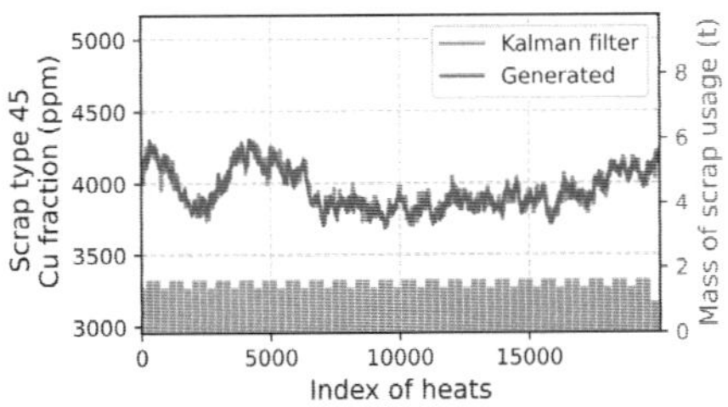


(a) Identity-based matrix
rank = 45, condition number = 1.

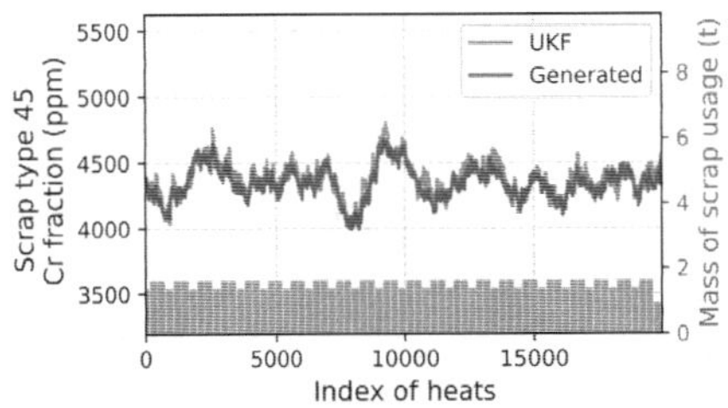


(b) Identity-based matrix
rank = 45, condition number = 1.

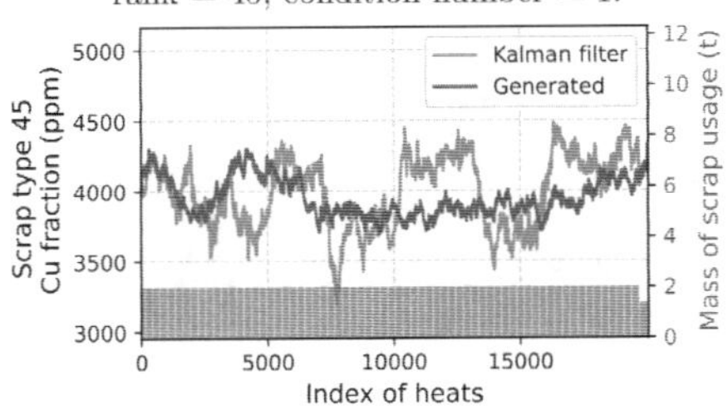


(c) Low rank matrix, rank = 20.

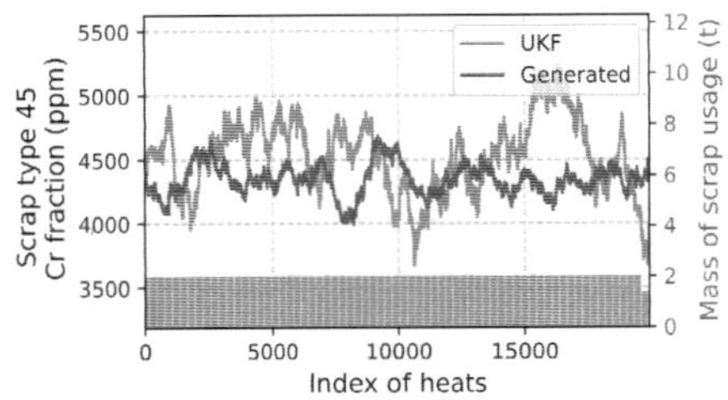


(d) Low rank matrix, rank = 20.

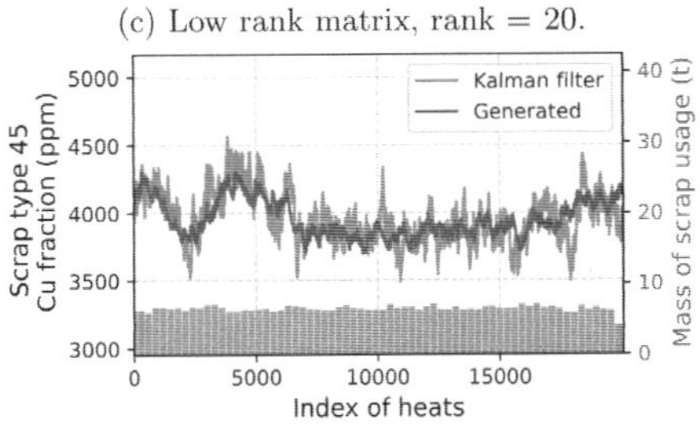


(e) Sparse matrix, rank = 45.

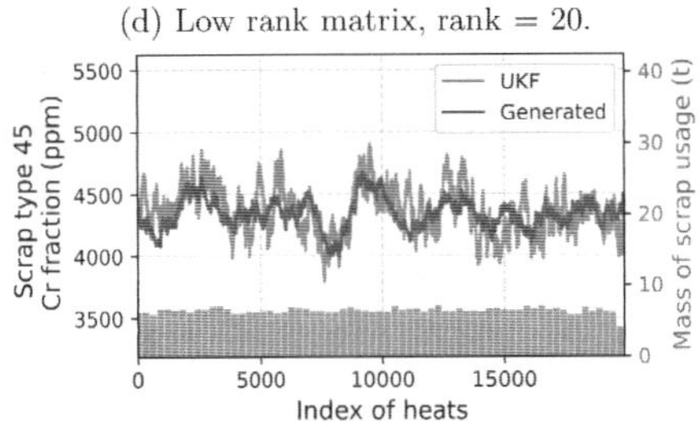


(f) Sparse matrix, rank = 45.

Figure 1: Effects of rank and condition number of matrix $M$ about input scrap mass on the estimated element fraction of scrap type 45. The left column is the results about Cu and the right column shows the results about Cr. The last row performs the best in estimating the scrap composition, since it is used frequently and with more quantities.

Table 3: Effect of measurement noise in scrap input mass on predicted Cu and Cr fractions in steel (ppm). The first row is the same as the second row in Table 2, representing the theoretical best-case results. Since the true value of $f_{\text{steel,Cu},t}$ range between 200 to 600 ppm, and $f_{\text{steel,Cr},t}$ ranges between 200 to 400 ppm, all results reported in the table are considered acceptable.

| noise ($\vec{m}_t$) | $\mathrm{E}[\varepsilon(f_{\text{steel,Cu},t})]$ | $\sigma[\varepsilon(f_{\text{steel,Cu},t})]$ | $\mathrm{E}[\varepsilon(f_{\text{steel,Cr},t})]$ | $\sigma[\varepsilon(f_{\text{steel,Cr},t})]$ |
|---|---|---|---|---|
| 0% | −0.33 | 10.79 | −0.07 | 5.39 |
| 1% | 0.23 | 10.47 | −0.12 | 5.33 |
| 5% | 0.32 | 13.30 | 0.44 | 7.13 |
| 10% | −1.52 | 16.69 | −0.18 | 9.31 |
| 20% | −3.21 | 19.25 | −1.26 | 12.88 |

Journal of Physics: Conference Series **3050** (2025) 012001 doi:10.1088/1742-6596/3050/1/012001

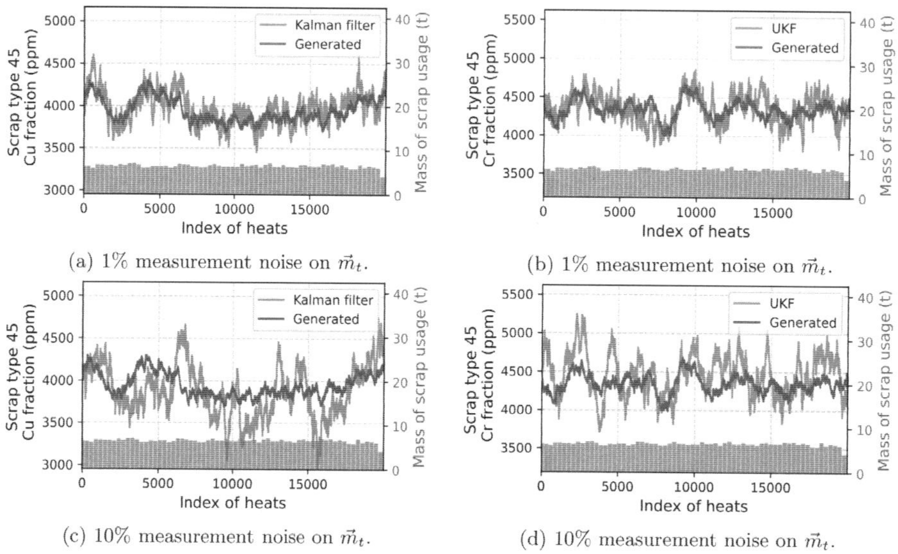


(a) 1% measurement noise on $\vec{m}_t$.
(b) 1% measurement noise on $\vec{m}_t$.
(c) 10% measurement noise on $\vec{m}_t$.
(d) 10% measurement noise on $\vec{m}_t$.

Figure 2: Effects of measurement noise on input scrap mass on the estimated Cu and Cr fraction of scrap type 45. The left column is the results about Cu and the right column shows the results about Cr. The red bars in the figure tell the usage of the scrap type.

This makes it important to understand how sensitive the model is to potential measurement errors in this quantity.

Table 4 and Figure 3 present the results of introducing controlled noise into the slag mass and evaluating its impact on the estimation of Cr content. Cu is not affected in this analysis, as slag-related variables are not part of the linear Cu model. The results show that the model maintains reasonable robustness to noise up to 20%, with only slight changes in bias and variance. However, the error of estimated Cr fraction of scrap gets larger when the noise level reaches 10%.

These findings highlight that among all the key process variables, the slag mass is one of the most influential in the nonlinear model. Accurate estimation of slag mass is thus essential for reliable prediction of steel composition and estimation of scrap composition. In industrial practice, improving the accuracy of slag mass estimation, through more direct measurements, better sensors, or refined empirical models, could significantly enhance the reliability of data-driven state estimation models like the one presented in this study.

Table 4: Effect of slag mass noise on predicted Cr fraction in steel (ppm). The first row is the same as the second row in Table 2, representing the theoretical best-case results. Since the true value of $f_{\text{steel,Cr},t}$ ranges between 200 to 400 ppm, all results reported in the table are considered acceptable.

| noise ($\vec{m}_{\text{slag}}$) | $\mathrm{E}[\varepsilon(f_{\text{steel,Cr},t})]$ | $\sigma[\varepsilon(f_{\text{steel,Cr},t})]$ |
|---|---|---|
| 0% | −0.07 | 5.39 |
| 1% | −0.01 | 5.44 |
| 5% | 0.52 | 7.21 |
| 10% | 0.12 | 9.21 |
| 20% | −0.07 | 13.06 |

### *3.4 Iron oxide fraction in slag*

The iron oxide fraction in slag, denoted as $f_{\text{FeOn,slag},t}$, is a key input for estimating the partition coefficient $\ell_X$, which governs the distribution of certain elements between steel and slag. In the model, this

Journal of Physics: Conference Series **3050** (2025) 012001 doi:10.1088/1742-6596/3050/1/012001

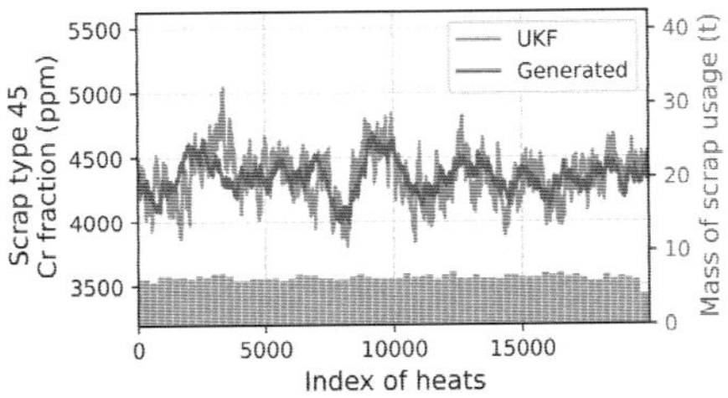


(a) 1% measurement noise on $m_{\mathrm{slag},t}$.

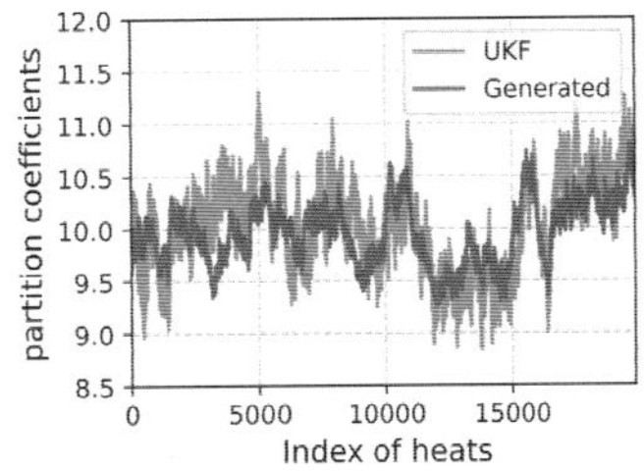


(b) 1% measurement noise on $m_{\mathrm{slag},t}$.

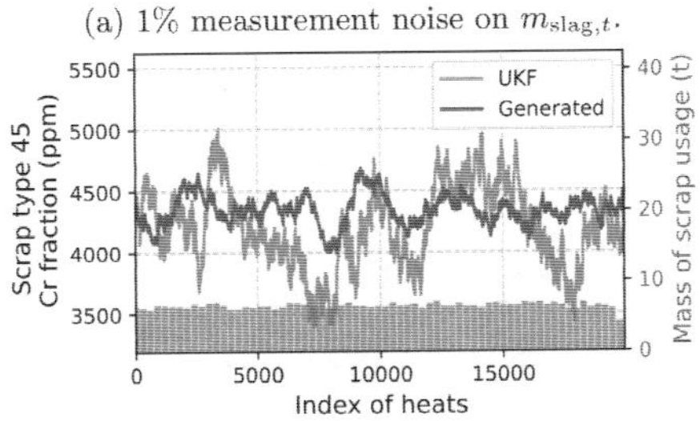


(c) 10% measurement noise on $m_{\mathrm{slag},t}$.

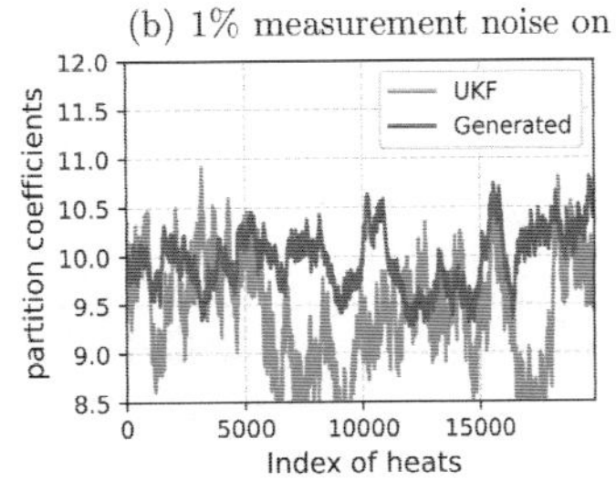


(d) 10% measurement noise on $m_{\mathrm{slag},t}$.

Figure 3: Effects of measurement noise on slag mass on the estimated Cr fraction of scrap type 80 and partition coefficients. The left columns are the results about Cr fraction of scrap type 45 and the right column shows the results about partition coefficients. The upper row shows the results with 1% measurements noise and the bottom row shows the results of 10% measurement noise of input scrap mass.

relationship is captured through a linear approximation of the form $\ell_t = c_{1,t} + c_{2,t} f_{\mathrm{FeOn,slag},t}$, where the parameters $c_{1,t}$ and $c_{2,t}$ are fitted coefficients. This was already successfully tested on real BOF data in [12].

Because this partition coefficient is modeled as a function of $f_{\mathrm{FeOn,slag},t}$, any error in measuring this variable could potentially propagate into the estimation of element concentrations. However, the use of a flexible linear parameterization allows the model to partially absorb or correct for such errors through the adjustment of $c_{1,t}$ and $c_{2,t}$. As a result, the model is expected to exhibit some robustness to moderate inaccuracies in this measurement.

This expectation is confirmed by the results shown in Figure 4, where the model is tested with 10% Gaussian noise in $f_{\mathrm{FeOn,slag},t}$. Despite this significant level of distortion, the estimated Cr fractions in scrap and the corresponding partition coefficients remain closely aligned with the true generated values. According to Table 5, both the bias and variance of the predictions remain low, and no significant deviation from the ground truth is observed.

These findings indicate that the current formulation for modeling the partition coefficient $\ell_X$ is sufficiently robust for practical applications. This robustness is particularly beneficial, as direct measurement of the iron oxide fraction in slag is often subject to uncertainty, relying on laboratory analysis or indirect estimates. The model's ability to tolerate such noise without compromising estimation accuracy enhances its applicability in real-world steelmaking environments, where measurement imperfections are unavoidable.

Overall, the results suggest that while accurate data is always preferable, the estimation of partition behavior through a simple linear model using noisy $f_{\mathrm{FeOn,slag},t}$ remains effective and stable. This reduces the dependency on high-precision slag composition measurements, potentially lowering operational constraints for real process monitoring and control.

## 4 Conclusion and Discussion

This study presented both linear and nonlinear state space models for estimating scrap composition in steelmaking, using mass balance equations relevant to Electric Arc Furnace (EAF) and Basic Oxygen Furnace (BOF) processes. A modified Kalman filter was applied for the linear model (with Cu as an example), and an Unscented Kalman Filter (UKF) was used for the nonlinear model (using Cr).

Table 5: Effect of measurement noise of iron oxide in slag on predicted Cr fraction in steel (ppm). The first row is the same as the second row in Table 2, representing the theoretical best-case results for this case. Since the true value of $f_{\text{steel,Cr},t}$ ranges between 200 to 400 ppm, all results reported in the table are considered acceptable.

| noise ($\vec{m}_{\text{slag}}$) | $\mathrm{E}[\varepsilon(f_{\text{steel,Cr},t})]$ | $\sigma[\varepsilon(f_{\text{steel,Cr},t})]$ |
|---|---|---|
| 0% | −0.01 | 5.39 |
| 5% | −0.30 | 5.07 |
| 10% | 0.03 | 5.29 |
| 20% | −0.32 | 5.34 |

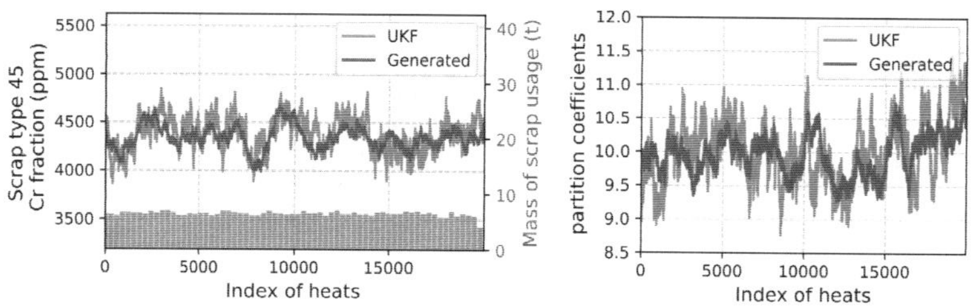


Figure 4: Effects of 10% measurement noise on iron oxide on the estimated partition coefficients and element fraction of scrap types. The left figure shows the estimated Cr fraction of scrap type 45 and the right figure shows the estimated partition coefficient.

The primary focus was to evaluate the sensitivity of these models to measurement noise in key process variables, including steel mass, steel composition, input mass of scrap types, slag mass, and iron oxide fraction in slag.

Key findings from the sensitivity analysis include:

- Errors in steel mass and composition can be effectively handled as observation noise.
- The model is relatively robust to noise in scrap input mass.
- The predicted elemental fraction in steel is quite robust to input mass matrices with different rank and condition number. However, accurate estimation of the elemental fraction of a specific scrap type is more likely when it is used more often and in greater amounts.
- The model is also sensitive to errors in slag mass, which can significantly affect the estimated fraction of scrap types.
- Errors in the iron oxide fraction have minimal impact, due to compensation by model parameters.

**Acknowledgment**
The research leading to these results has been performed within the CAESAR project (https://caesarproject.eu/) and received funding from the European Commission's Horizon Europe Programme under grant agreement n° 101058520.

**References**

[1] Niloofar Arzpeyma, Moudud Alam, Rutger Gyllenram, and Pär G Jönsson. Model development to study uncertainties in electric arc furnace plants to improve their economic and environmental performance. *Metals*, 11(6):892, 2021.

[2] Jean Pierre Birat, Philippe Russo, et al. *Quality of heavy market scrap – Development of new and simple methods for quality assessment and quality improvement – Final report.* Publications Office of the European Commission, 2002.

[3] Lotte De Vos, Inge Bellemans, Carina Vercruyssen, and Kim Verbeken. Basic oxygen furnace: assessment of recent physicochemical models. *Metallurgical and Materials Transactions B*, 50:2647–2666, 2019.

[4] James Durbin and Siem Jan Koopman. *Time series analysis by state space methods*, volume 38. OUP Oxford, 2012.

[5] Gabrielle Gaustad, Preston Li, and Randolph Kirchain. Modeling methods for managing raw material compositional uncertainty in alloy production. *Resources, Conservation and Recycling*, 52(2):180–207, 2007.

[6] Thomas Hay, Ville-Valtteri Visuri, Matti Aula, and Thomas Echterhof. A review of mathematical process models for the electric arc furnace process. *Steel Research International*, 92(3):2000395, 2021.

[7] Simon Julier and Jeffrey Uhlmann. New extension of the Kalman filter to nonlinear systems. In *Signal Processing, Sensor Fusion, and Target Recognition VI*, volume 3068, pages 182–193. Spie, 1997.

[8] Simon Julier and Jeffrey Uhlmann. Unscented filtering and nonlinear estimation. *Proceedings of the IEEE*, 92(3):401–422, 2004.

[9] Rudolph Emil Kalman. A new approach to linear filtering and prediction problems. *Transactions of the ASME–Journal of Basic Engineering*, 82(Series D):35–45, 1960.

[10] Todd Moon. The expectation-maximization algorithm. *IEEE Signal processing magazine*, 13(6):47–60, 1996.

[11] Erik Sandberg, Barry Lennox, and Patrik Undvall. Scrap management by statistical evaluation of EAF process data. *Control Engineering Practice*, 15(9):1063–1075, 2007.

[12] Yiqing Zhou, Karsten Naert, and Dirk Nuyens. Modeling scrap composition in electric arc and basic oxygen furnaces. *arXiv preprint arXiv:2504.09382*, 2025.

# Application of a neural network for calculating the off-gas generation within an EAF under modified operating conditions

**Alexander Reinicke*,[1], Lilly Schulte[1] and Thomas Echterhof[1]**

[1] Department of Industrial Furnaces and Heat Engineering, RWTH Aachen, Germany

*E-mail: reinicke@iob.rwth-aachen.de

**Abstract.** By design, the electric arc furnace (EAF) emits low carbon dioxide emissions when compared to other aggregates such as the blast furnace. For this reason, electric steelmaking is at the focus of the current industrial transformation towards a $CO_2$ neutral steel production. In this context, the EAF is subject to further optimization of the energy and resource efficiency. Process models can be used to simulate and evaluate altered operating conditions within the EAF before conducting costly plant trails. With regards to these models, the off-gas generation plays a crucial role, as it directly affects both the carbon footprint of the process as well as the process conditions. Unfortunately, an off-gas analysis is often unavailable or malfunctioning, as the equipment is exposed to extreme conditions. For this reason, a model of the off-gas generation was developed based on an artificial neural network. The model utilizes only a-priori process data. This way the model can be used to complement an installed off-gas analysis in the event of a malfunction or be integrated into an existing EAF process model in order to consider the variation in off-gas generation when optimizing the furnace operating chart. The model is trained and tested on an industrial scale EAF. In addition, the original dataset is complemented by a parametrizable data generation module, with which the time period of the utilized process data can be adjusted. This way, training of the model is more independent of plant characteristics such as delays in measurement or reaction time within the melt or gas phase. This method is intended to improve the transferability of the results to different plants, in comparison to, for example, recurrent neural networks. The developed model can predict the off-gas mass flow over the duration of a single heat in good agreement, achieving an average absolute error of approximately 8 percent. Larger deviations can be partially attributed to previously unidentified malfunctions in the measurement equipment. During post-processing, multiple indicators are calculated and a process model based on physical principles is applied to identify any conflicting results and correct them if applicable.

## 1. Introduction

The electric arc furnace (EAF) is characterized by its high flexibility regarding the charged ferrous material. For this reason, it is primarily utilized for melting steel scrap. In contrast to the blast furnace route using ore-based material, recycling of steel scrap in an electric arc furnace (EAF) saves large amounts of $CO_2$ and therefore significantly contributes to the goal of a circular economy [1]. Recent studies have also shown that natural gas usage within the EAF can be substituted with hydrogen further reducing the need for fossil fuels [2]. Lastly with the shift towards production of steel from directly reduced iron (DRI), the EAF may become the center piece of steel production, replacing the blast furnace.

At the same time, the changes pose challenges to the established operating procedures in the EAF. These include maintaining high productivity and resource efficiency while minimizing, for example, the wear of the refractory material [3]. In order to achieve these goals, studies and plant trial must be performed to investigate the influence of the changed operating conditions on the plant's performance indicators.

In general, mathematical models are often used for simulation and evaluation of industrial processes as they allow investigation of changes before conducting costly experiments. Hay et al. [4] published a comprehensive overview of the available EAF models. These models can be distinguished based on their approach in calculating the off-gas composition. MacRosty and Swarz [5], Arnout et al. [6] as well as Kho et al. [7] apply the Gibbs free energy minimization to determine the chemical equilibrium. In contrast, Logar et al. [8,9] use the equilibrium constant method. Additionally, the models differ in their representation of the furnace geometry, reaction areas and the corresponding reaction kinetics.

Based on the work by Logar a comprehensive process model of the EAF was developed by Meier [10] and Hay [11]. The model is continuously validated using data of multiple industrial scale EAFs and is featured in multiple publications [12]. In addition to the simulation of historical heats, the model will be expanded to optimize the operation of the furnace regarding the usage of chemical and electrical energy. For this purpose, the operating chart is discretized with respect to the progression of the heat (e.g., time). The operating chart can then be automatically adjusted and used to simulate proposed changes. The results are evaluated based on several criteria including the overall usage of electrical energy as well as oxygen, natural gas and carbon.

The off-gas mass flow rate has previously been considered as an input to the EAF process model. The mass flow extracted through the off-gas duct creates negative pressure inside the EAF. The pressure difference between the environment and the interior results in ambient air entering the furnace. This provides additional oxygen for chemical reactions in the gas phase. However, nitrogen is also introduced into the EAF. The nitrogen hardly reacts within the EAF but is heated and thus carries energy out of the process. Adjustment of the furnace operation, particularly regarding the use of oxygen and natural gas by the burners and lances, changes the off-gas mass flow generated in the furnace. However, the furnace control often operates independently of this, which ultimately results in a variation in the amount of leak air entering the EAF. Therefore, the variable mass flow rate must be considered in the simulation of the heat as well as in the evaluation of the furnace emissions. For this reason, a two-stage model for approximation of the off-gas mass flow rates was developed. In the first stage the off-gas mass flow rate is estimated using a neural network. In the second step, the calculated mass flow rate is iteratively adjusted by using the process model and evaluating the atmospheric pressure inside the EAF as calculated by the process model.

## 2. Method

Natural gas (mainly methane) and oxygen are fed into the furnace through the lances and burners. Depending on the gas supply, the oxygen used contains a few percent of nitrogen and argon [13]. In addition, ambient air is regularly drawn into the EAF. In the following, the ambient air is assumed to be a mixture of nitrogen, oxygen and argon. The variable mass fraction of water vapor in the ambient air is unknown and neglected. The composition of the supplied gases is summarized in **Table 1**.

**Table 1**. Composition of the injected gases

| | Composition in wt.-% | | | |
|---|---|---|---|---|
| Species | $CH_4$ | $O_2$ | $N_2$ | Ar |
| Natural gas | 99 | 0 | 1 | 0 |
| Oxygen | 0 | 94 | 3 | 3 |
| Ambient air* | 0 | 23.2 | 75.5 | 1.3 |

*Water vapor content of ambient air is neglected

In addition to the species already mentioned, the off-gas also contains carbon monoxide and carbon dioxide. Carbon dioxide and water vapor are formed as reaction products during the combustion of natural gas according to Equation 1. In the context of this work, stoichiometric combustion and an almost complete conversion of methane are assumed at all times. Water vapor is also released from the moisture contained in the ferrous material and the water from the electrode cooling. In addition, carbon dioxide can be formed by the post-combustion of carbon monoxide with oxygen in the furnace atmosphere according to Equation 2. Carbon monoxide is mainly produced within the EAF by the reaction of injected carbon and oxygen as stated in Equation 3. The carbon is present in elemental form (charged with a basket or blown in) or is dissolved in the melt.

$$CH_4 + 2\,O_2 \rightarrow CO_2 + 2\,H_2O \tag{1}$$

$$CO + \frac{1}{2}O_2 \leftrightarrow CO_2 \tag{2}$$

$$C(s) + \frac{1}{2}O_2 \rightarrow CO \tag{3}$$

The off-gas composition of the investigated furnace is determined using a mass spectrometer. For this purpose, part of the off-gas is extracted from the off-gas duct. The off-gas is cleaned and cooled before it reaches the spectrometer. Most of the moisture in the off-gas is removed during the pre-processing, therefore the analysis values are converted to the values for dry off-gas [14]. The off-gas mass flow can then be determined using the known composition of injected oxygen and ambient air using a system of equations according to Equation 4 and Equation 5 [15]. The absorption of nitrogen into the melt can be disregarded due to the comparatively small quantities, typically in the range of 100 ppm [16,17]. The total mass flow of the leak air is also calculated using the neural network. However, the quantity of leak air that is injected downstream of the EAF at the slip gap cannot be determined using this method.

$$\dot{m}_{Oxygen} \cdot w_{N_2,Oxygen} + \dot{m}_{Leak\ air} \cdot w_{N_2,Leak\ air} = \dot{m}_{Offgas} \cdot w_{N_2,Offgas} \tag{4}$$

$$\dot{m}_{Oxygen} \cdot w_{Ar,Oxygen} + \dot{m}_{Leak\ air} \cdot w_{Ar,Leak\ air} = \dot{m}_{Offgas} \cdot w_{Ar,Offgas} \tag{5}$$

The pre-processing of the extracted off-gas and the analysis itself are associated with a time delay. The same applies to the flow measurement of the injected gas and coal in reverse direction. The average delays were determined in advance based on a statistical analysis of the process data. The data series are shifted by the respective delays before further processing. **Figure 1** shows an example of the calculated off-gas mass flow for a single heat. The mass flow and the process time are normalized to the highest value within the heat. In addition, the cumulative mass flow of oxygen and natural gas as well as the mass of injected carbon are shown in a staggered manner.

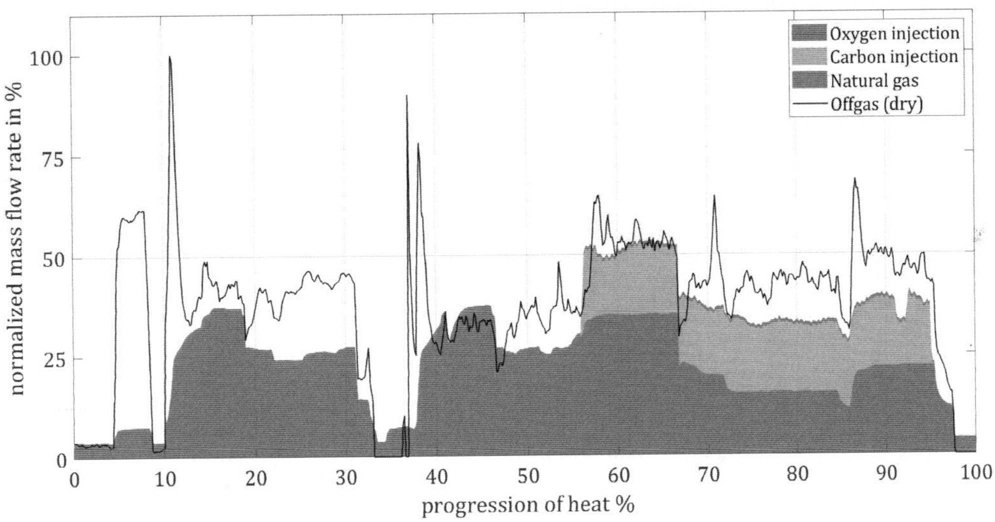


**Figure 1.** Comparison of in- and outgoing mass flow rates

The graph illustrates that the injected and discharged mass flow mismatches in some stages of the process. There are several reasons for the difference in mass. During scrap charging at approximately 20% and 45% progression of the heat the furnace roof is moved to the side and the off-gas is not directed into the off-gas duct but to a large extent escapes into the surrounding hall. At the same time, a high mass flow of ambient air is sucked into the off-gas duct. As a result, the measured off-gas composition is corrupted and the solution of the system of equations does not provide a meaningful result. Scrap charging must therefore be excluded from the evaluation and training of the model. In addition, significant differences also occur in the later stages of the heat. These can be mainly attributed to leak air being sucked into the EAF. In addition, carbon and oxygen are also absorbed by the melt or bound in the form of oxides in the slag or released back into the furnace atmosphere. The reaction rate and the residence time in the furnace are highly dependent on the current state of the process. Leak air intake is also determined by the general design and the current power setting of the off-gas extraction system. The analysis shows that the off-gas mass flow cannot be easily determined by summing of the injection rates. When investigating alternative operating modes, measurement data of the off-gas composition is however not available.

For this reason, the off-gas mass flow rate is estimated by using a two-stage model as shown in **Figure 2**. In the first step, an initial solution for the mass flow is calculated by application of a multi-layer feedforward neural network and using the a-priori process data. The data preparation and training of the neural network is described in detail in the following section. The non-linear nature of the model enables the representation of complex interactions between the different variables of the operating chart [18]. In the second step, the heat is simulated by applying the EAF process model and using the predicted off-gas mass flow as an input. In this regard, the objective is to adjust the estimated off-gas mass flow rate according to the pressure within the EAF. In contrast to the purely statistical nature of the neural network, the process model is based on the principles of energy and mass conservation [10,11]. The calculation of the pressure is subject to the intake of leak air and the extraction of off-gas. The leak air intake in turn is determined using the furnace pressure. The process model is parameterized and validated using historical production data, therefore providing a suitable reference for the results of the neural network.

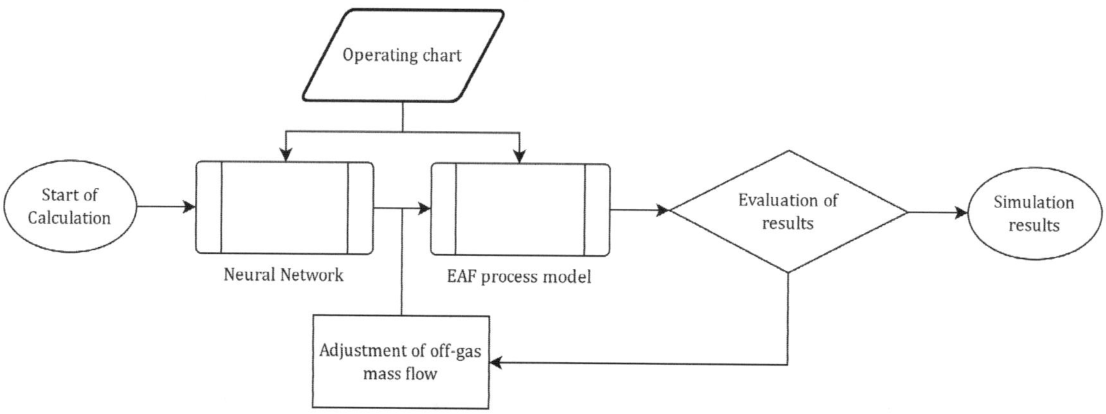


**Figure 2**. Flow chart of the off-gas approximation

The off-gas extraction creates a negative pressure in the range of a few millibars, depending on the stage of the process. Any deviation from this may imply an over- or underestimation of the off-gas mass flow rate, which may also result in an unstable operation of the process model. The mass flow rate is corrected by evaluating the difference in pressure as well as the pressure gradient according to Equation 6 and then applying the correction coefficient to reduce or increase the extracted off-gas mass flow. In the scope of this work the target pressure was set at -3 Pa. In addition, the target pressure was adjusted for the early phase of scrap meltdown as well as carbon injection. It is assumed that the furnace pressure is higher during these stages due to the gasification of the volatile components in the scrap and the formation of CO during slag foaming, respectively. The adjustment of the mass flow rate and the simulation using the process model are repeated until the model is stable and the calculated furnace pressure is within a realistic range. The main objective of this method is to create valid off-gas mass flow rates for further application of the process model.

$$f_{Corr,Pressure} = C_1 \cdot (p_{calc} - p_{target}) + C_2 \cdot \frac{dp}{dt} \quad (6)$$

The structure of the implemented neural network is depicted schematically in **Figure 3a**. The network consists of an input and output layer as well as three intermediate layers with 40, 30 and 20 neurons respectively. The input values include the furnace operation chart and the mass of the charged input materials. The utilized data is summarized in **Table 2**. It includes the current values for the input of electrical and chemical energy as well as the corresponding integrals. Moreover, the percentages of electrical energy input and time in relation to the end of the basket have been added. The identification number and timestamp of each heat are excluded from the training data. Furthermore, the dataset is shuffled in order to avoid overfitting of the neural network to specific heats, production campaigns or seasonal effects. However, each heat can ultimately be identified from a combination of its characteristics such as the mass of charged scrap and sum of injected oxygen and carbon or its overall electrical energy input. Nevertheless, the information utilized is important for estimation of the off-gas generation from a metallurgical point of view.

**Table 2**. Overview of the input data of the neural network

| Time and electrical Input | Operating Chart Injection of: | Integrals | Relative Measures | Charged Material |
|---|---|---|---|---|
| Elapsed time of current heat | Natural gas | Tap-to-Tap Time | Electrical Energy | Total mass in basket |
| Elapsed time of current basket | Oxygen (Burner) | Oxygen Lance | Basket Time | Water content of scrap* |
| Basket number | Oxygen (Lance) | Carbon Injection | | Volatile components of scrap* |
| Electrical Power | Oxygen (Post combustion) | | | Carbon content of scrap* |
| Voltage | Carbon | | | Charged carbon (basket) |
| Electrical Current | Dust | | | |

*Calculated based on scrap mass and chemical composition of samples

In addition, the integrals of the continuous process variables are used to describe the influence of the previous operation on the current state of the EAF. The discrete values of the operating diagram are first multiplied by a weighting function and then summed over the observation period. The weighting is based on a normal distribution with a parameterizable mean value, standard deviation and a left-sided cut-off. **Figure 3b** shows an example of the use of oxygen lances and the applied weighting function. The horizontal axis depicts the distance to the time of observation. The parameterizable normal distribution stated in Equation 8 was chosen to represent the different reaction times within the arc furnace as well as the delays caused by the measurement equipment used. In this context, the mean value (μ) describes the average time between the cause and the peak in amplitude. It is influenced by the gas velocity and the distance between, for example, the burners and the sample extraction within the off-gas duct. The standard deviation ($\sigma^2$) on the other hand is affected by the volume of the furnace. In theory within a larger vessel, the change in the mass flow rate is smoother and more leveled. Lastly, the cut-off defines the maximum period under consideration.

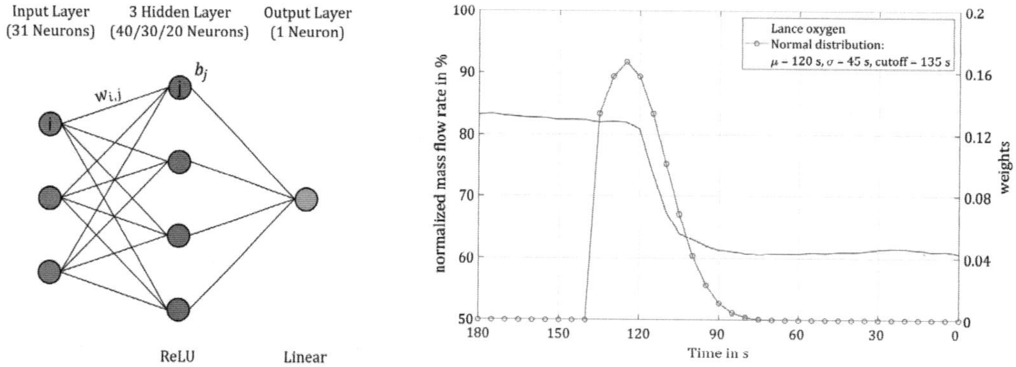


**Figure 3.** a) Structure of the neural network (left); b) application of the normal distribution (right)

$$f(x,\mu,\sigma^2,cutoff) = \frac{\frac{1}{\sqrt{2\pi\sigma^2}} e^{-\frac{(x-\mu)^2}{2\sigma^2}}}{\sum_{i=cutoff}^{0}\left(\frac{1}{\sqrt{2\pi\sigma^2}} e^{-\frac{(i-\mu)^2}{2\sigma^2}}\right)} \tag{8}$$

The calculation of the output values of individual neurons is stated in Equation 9. The input values or the output of the previous layer $v_j$ are multiplied in each node $i$ by weighting factors $w_{ij}$. Afterwards a bias $b_i$ is added before an activation function is applied, determining the final output value of the neuron.

$$x = \sum_{j=1}^{n}(w_{ij}v_j) + b_i \tag{9}$$

The hidden layers are processed using the Rectified Linear Unit (ReLU), while a linear activation function is used in the output layer. The respective calculation rules are given in Equation 10 and Equation 11. Both functions show a linear increase in the output value. In contrast to the linear activation function, the output value of the ReLU is however limited to 0. This is achieved by averaging with the absolute value of the input. The averaging adds a non-linear element to the neural network, which is crucial for multi-layer networks, as the layers can otherwise be reduced to a single layer using linear combination [18,19]. Application of the ReLU therefore allows for representation of more complex systems.

$$f_{linear}(x) = a \cdot x \tag{10}$$

$$f_{ReLU}(x) = \frac{x + |x|}{2} \tag{11}$$

The described method was deliberately chosen over the application of a more sophisticated type of neural network (e.g., recurrent neural network or LSTM) in order to separate the temporal component from the actual prediction. In this way, the model is aimed to retain limited usability even when applied to different furnaces. Once trained, a recurrent neural network projects the characteristics of the original dataset (e.g., delays and reaction times) onto the new data used for prediction. In order to modify the behavior of the model, the weights of the network must be retrained. Adjusting the parameters of the normal distribution allows the model to be altered without retraining. However, calibrating these parameters using solely the geometry of the furnace necessitates further investigation of possibly multiple furnaces.

To complement the investigation, a LSTM model was implemented to verify whether the preprocessing step is able to partially substitute the predictive capabilities of more complex model types. The overall structure of the LSTM model (long short-term memory) is largely the same as the neural network shown in **Figure 3a.** However, the hidden layers have been substituted with two consecutive layers, each consisting of 22 LSTM nodes. LSTMs are a subtype of recurrent neural networks, meaning the hidden state of each node is also used as an input in the next iteration. The resulting feedback loop is depicted schematically in **Figure 4a**. Furthermore, LSTMs utilize an additional internal cell state, that acts as a memory. The composition of an individual LSTM node is shown in **Figure 4b**. It is subdivided into three functional components (gates), that control whether information in the cell state is erased, added or utilized in the node's output. In this context, the sigmoid and hyperbolic tangent (tanh) provide differentiable nonlinear activation functions that map the input to output values between 1 and 0 (or -1, respectively). Aggarwal [18] and Goodfellow et al. [20] provide a more detailed description of the functionality, benefits as well as limitations of the different types of recurrent neural network.

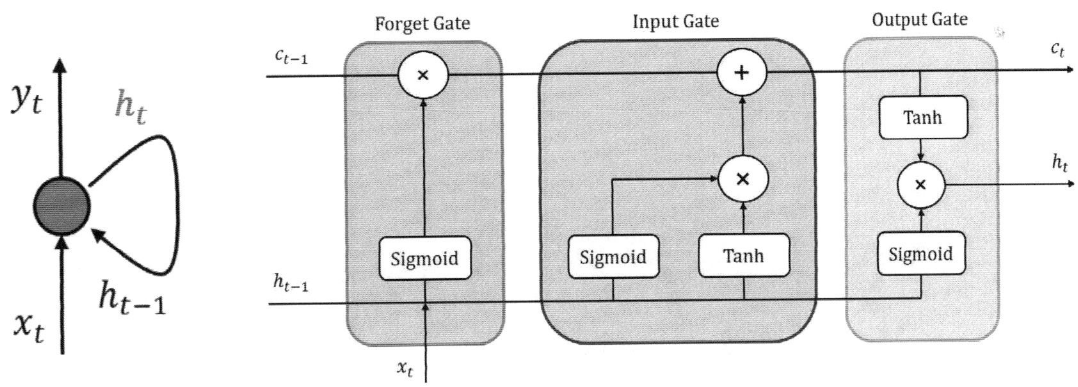


**Figure 4.** a) Feedback loop within an individual node of a recurrent neural network (left); b) Functional components of a LSTM node (right)

Training of the neural network was carried out using the process data of about 10.000 heats spanning the production of two years. A validation split of 30% was used for evaluation of the model results. The process data was first divided into overlapping segment with a fixed length of 70 seconds. Afterwards, data segments with faulty or inconsistent data were removed based on multiple criteria. A key objective was to exclude segments in which mainly leak air was vented through the exhaust duct. These can be identified by a high percentage of both nitrogen and oxygen as well as a comparatively low temperature. Moreover, unreasonably high and low mass flow rates were excluded as well, as they are likely a consequence of corrupted composition measurements, as previously stated.

Journal of Physics: Conference Series **3050** (2025) 012002 doi:10.1088/1742-6596/3050/1/012002

Conditions for removal of data segments:

- Missing values for the off-gas mass flow rate or any of the used input data
- Measured mass flow rate above or below threshold
- Nitrogen content of off-gas above 50-wt.%
- Oxygen content in off-gas above 20-wt.%
- Off-gas temperature below 1000°C

The results of the model are evaluated based on the average deviation between the prediction and the measurements as stated in Equation 12. In this regard, $f_i$ denotes the prediction of the neural network. The measured mass flow rate is given by $y_i$.

$$MAE = \frac{1}{n}\sum |(f_i - y_i)| \tag{12}$$

## 3. Results

The estimated off-gas mass flow rates calculated by the neural network in relation to the measurement are presented in **Figure 5**. The plot contains the entire dataset of roughly 10.000 heats, totaling about 3.4 million data points. For 53% of these at least one of the previously mentioned conditions is met, meaning they would have been removed during training of the model. Regarding the data points that fulfill all criteria, the model on average deviates from the measurement by about 8% of the maximum mass flow rate. However, the graph indicates that larger deviations of up to 50% are not uncommon.

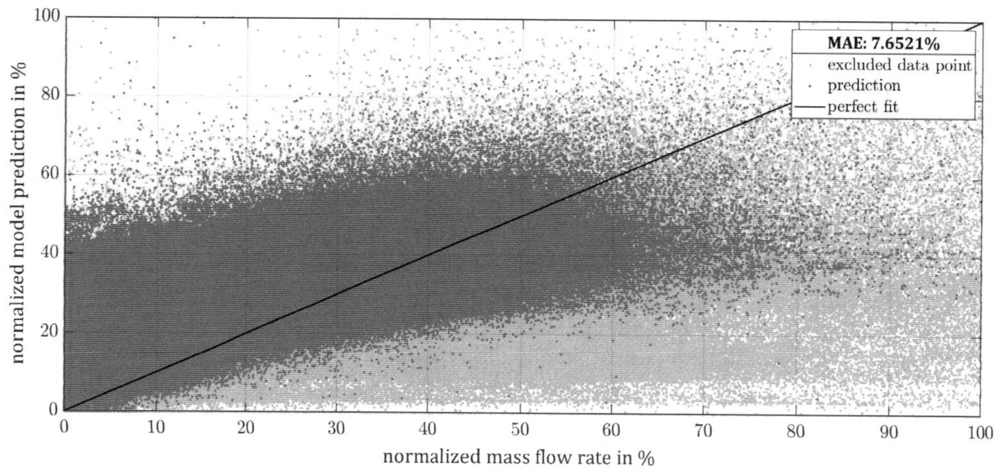


**Figure 5**. Normalized mass flow rate calculated using the neural network compared to measurements

In **Table 3**, the results of the neural network when excluding and including the composed integrals are compared to the results of the LSTM. Additionally, the number of trainable model parameters is provided. The best results are obtained by either including the calculated integrals or using the LSTM, with only a minor difference between both models. When omitting the integrals, the mean absolute error increases by about one percent. This suggests that both approaches can reproduce the delay between the injection of mass into the furnace and the subsequent generation and removal of the off-gas.

The feedforward network contains approximately 3200 trainable parameters. Considering the dataset used for training, this is about three heats for each parameter. However, the off-gas generation arguably depends not only on the input material, but also on the operating condition of the furnace. Therefore, time segments of the same heat can be considered distinct samples if the operating conditions are significantly different. In this regard, the EAF process can be divided into multiple stages. These include the bore-down of the electrodes, the subsequent meltdown of the scrap using high electric power, and finally the flat-bath phase and slag foaming [21]. Nevertheless, the model may still be subject to overfitting, necessitating the need for further validation on future heats.

**Table 3**. Comparison of the model results and the number of trainable parameters

| Model | Neural Network excluding integrals | Neural network including integrals | Long Short-Term Memory |
|---|---|---|---|
| Relative MAE* | 8.77% | 7.65% | 7.75% |
| Number of trainable parameters | 2791 | 3191 | 7943 |

*Mean absolute error divided by the maximum mass flow rate

The prediction of the feedforward neural network for two exemplary heats is shown in **Figure 6** and **Figure 7**. In addition, the results obtained by the LSTM model are displayed. Contrary to the overall results, the models are able to predict the mass flow rate of the first heat in good agreement. At the beginning of the second basket, between 40% and 50% of the heats progress, the model matches the measurement even though the segment is flagged for exceeding the range of training data. Over the duration of the heat there are only minor deviation between the feedforward network and the LSTM. However, the measured mass flow rate generally displays greater fluctuation than the calculated values.

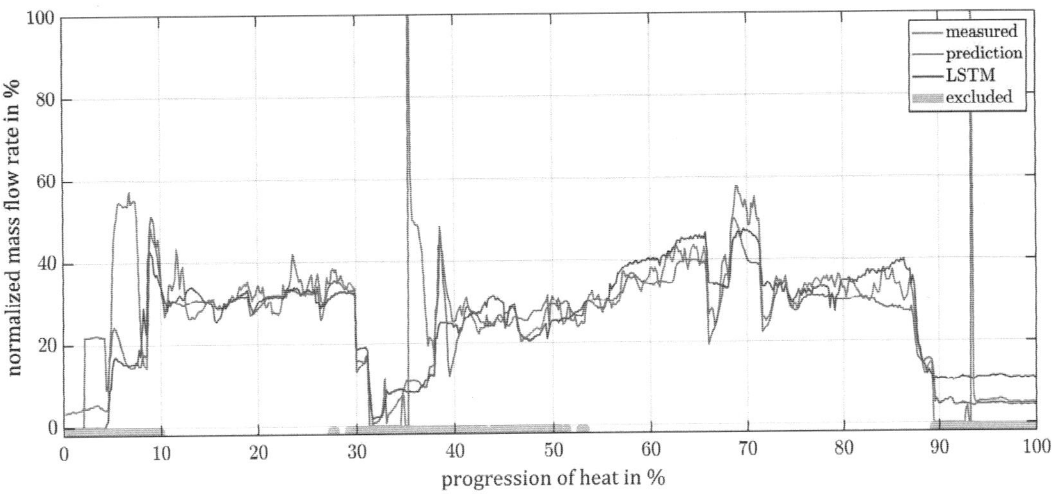


**Figure 6**. Comparison of the calculated and measured off-gas mass flow rate over the duration of heat 1

In contrast to this, for the second heat both models overestimate the off-gas generation over the entire duration of the heat. At the same time the measured mass flow rate is also significantly lower than in the first case. Following the second heat, the off-gas measurement becomes increasingly unstable until it finally fails. This leads to the assumption that the off-gas measurement of the depicted heat was already faulty to some extent. Moreover, in **Figure 8** the measured mass flow rate of the second heat is contrasted by the injection of oxygen, natural gas and carbon. The operating chart is similar to the mass flow rates displayed in **Figure 1**. However, the injected mass exceeds the extracted off-gas over the entire duration of the heat, further indicating a faulty measurement. Considering the previous process data, it is rather unlikely that more than 50% of the injected oxygen will react with the melt to form oxides.

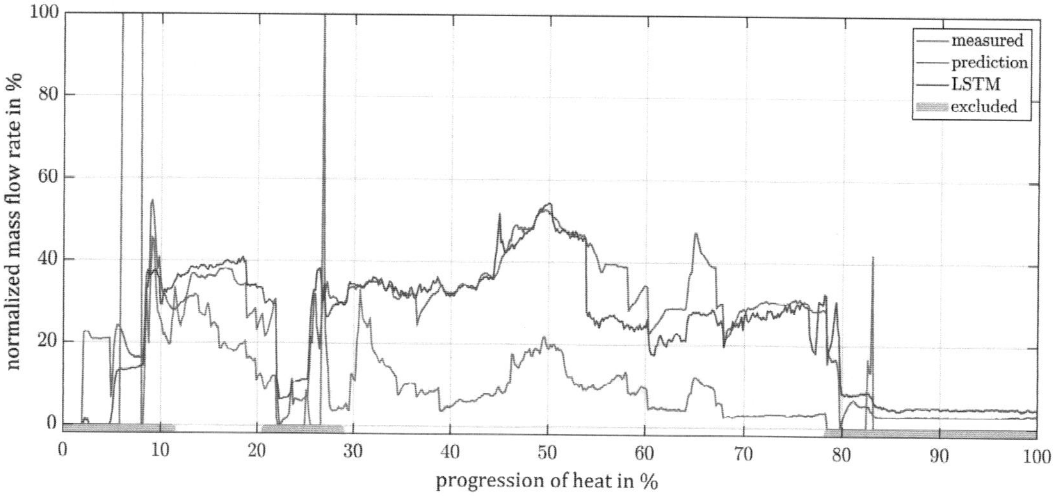


**Figure 7**. Comparison of the calculated and measured off-gas mass flow rate over the duration of heat 2

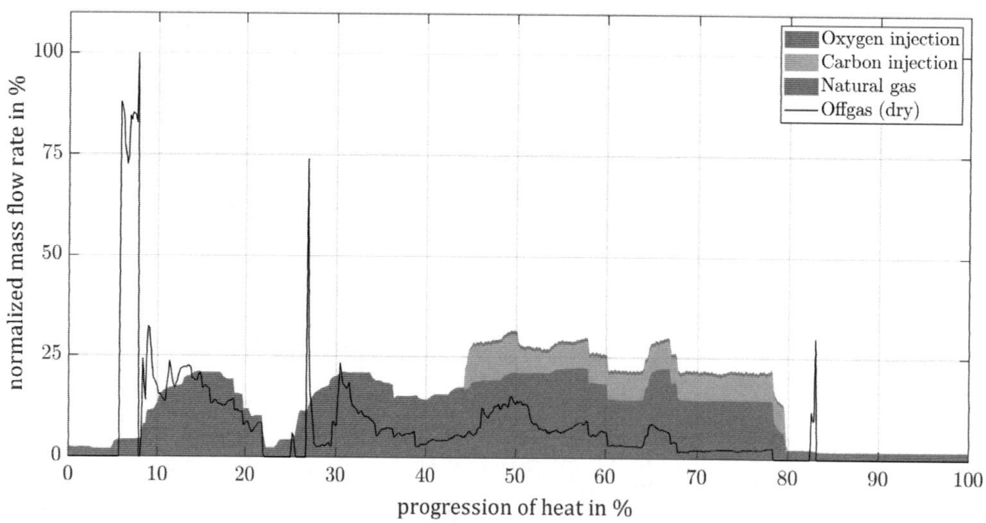


**Figure 8**. Comparison of in- and outgoing mass flow rates for heat 2

In **Figure 9** another example of a faulty off-gas measurement is provided. Over the duration of the heat the off-gas is composed of mostly nitrogen. This leads to the calculation of abnormally high mass flow rates. In contrast, the model still gives reasonable results but flags the results as potentially anomalous.

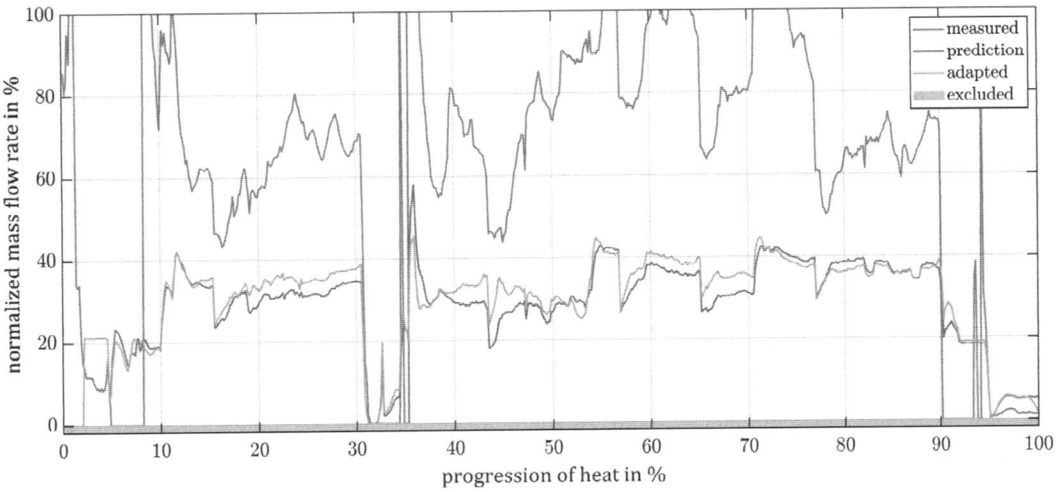


**Figure 9**. Comparison of the calculated and measured off-gas mass flow rate over the duration of heat 3

Lastly, the results of the pressure-based correction method for another heat are shown in **Figure 10**. The corresponding pressure for the original prediction of the neural network as well as after application of the correction step are displayed in **Figure 11**. During the meltdown of both scrap baskets the pressure decreases, as the gasification of the volatile components subsides. With the injection of carbon and oxygen in the flat bath phase, the pressure then peaks at over 20 Pa for the original prediction. As a result, the correction method increases the off-gas mass flow rate during slag foaming towards the end of the heat. This improves the prediction of the off-gas when compared to the initial estimate. However, the calculation of the pressure within the furnace is subject to a large number of inputs, ranging from the consumption of chemical and electrical energy and the temperature of the off-gas to the general geometry of the EAF and the rate at which ambient air is drawn into the furnace. At the same time, no direct measurement of the pressure within the furnace is available. The target pressure was derived entirely from the results of the process model using the measured off-gas mass flow rate and process data of the preceding production. It is therefore susceptible to modelling errors.

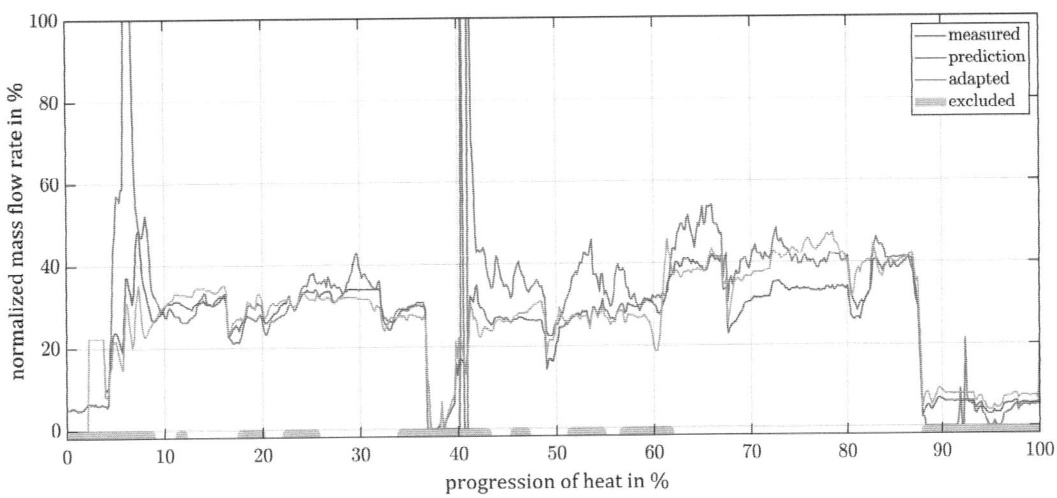


**Figure 10**. Application of the correction algorithm on the results of heat 4

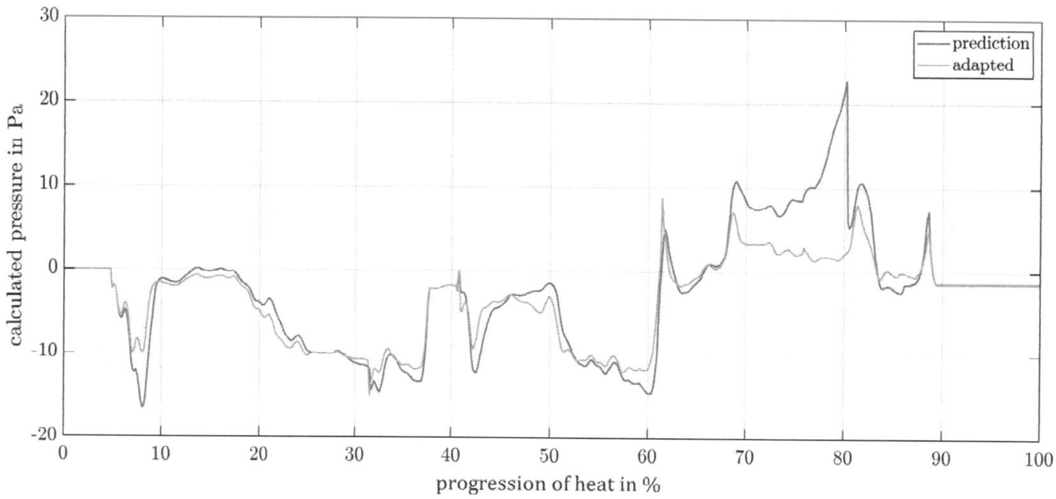


**Figure 11**. Comparison of calculated furnace pressure of heat 4

Directly measuring the atmospheric pressure within the EAF is difficult due to the extreme conditions inside the furnace. However, a camera-based approach could be used for future validation. Although a camera cannot provide quantitative measurements, the ejection of gas from the furnace indicates an overpressure in the EAF. The occurrences of these events can be compared to the pressure calculated using the process model.

## 4. Discussion

Within this work a neural network for estimation of the off-gas generation of an EAF was developed. The model shows a moderate accuracy with a mean absolute deviation of about 8% when compared to the measured off-gas mass flow rate. It is able to predict the overall magnitude of the off-gas emission and can represent key events such as the coal injection during the flat bath phase. Although about half of the data was removed from the evaluation, it was shown that the large deviation of the model results may be in part due to faulty measurements. Therefore, the pre-processing of the training data needs to be further refined. This can be accomplished by, for example, evaluating of the in- and outgoing mass flow rates over the entire duration of the process.

The remaining deviation between the model results and the measurements can be attributed to further reasons. The EAF process is a highly dynamic process and sampling of the process data (including the off-gas measurements) is limited to five second intervals. Abrupt events like collapse of the scrap during the meltdown might not be captured. The collapsing scrap leads to a pressure surge causing a large discharge of off-gas and subsequent intake of leak air. Furthermore, the measurement of the off-gas as well as the composition of the charged material and the injected gases are also subject to error. The off-gas composition is only measured after injection of additional leak air at the slip gap. Finally, relevant information on the state of the furnace might not be recorded at all. A possible example is the state of the slag door. When open, the intake of ambient air in the vicinity of the slag door is significantly increased.

However, as shown in the exemplary heats, the developed model can provide a good estimate of the off-gas mass flow even when exceeding the range of data used for training. As such the model can be utilized as a failsafe or monitoring tool for the equipment installed at the EAF. The pressure-based correction method is capable of generating mass flow rates suitable for stable simulation of the EAF process. On the other hand, the pressure calculated within the process model is subject to a large number of influences and is therefore susceptible to modelling errors. Validation of the off-gas prediction on modified operating charts has not yet been performed.

For future development of the approach, the training of the model should also be extended to additional furnaces, possibly featuring different operating conditions, such as charging of large quantities of DRI. This will allow the application of the weighted averaging of the process data and the pressure-based correction to be further tested and improved.

## References

[1] IEA. Iron and Steel Technology Roadmap. Available online: https://www.iea.org/reports/iron-and-steel-technology-roadmap.

[2] Schüttensack, L.; Reinicke, A.; Echterhof, T. Application of hydrogen operated burners in the electric arc furnace. IOP Conference Series: Materials Science and Engineering 2024, 1309, 12011, doi:10.1088/1757-899X/1309/1/012011.

[3] Heo, J.; Park, J.H. Interfacial reactions between magnesia refractory and electric arc furnace (EAF) slag with use of direct reduced iron (DRI) as raw material. Ceramics International 2022, 48, 4526–4538, doi:10.1016/j.ceramint.2021.10.238.

[4] Hay, T.; Visuri, V.-V.; Aula, M.; Echterhof, T. A Review of Mathematical Process Models for the Electric Arc Furnace Process. steel research int. 2021, 92, 2000395, doi:10.1002/srin.202000395.

[5] MacRosty, R.D.M.; Swartz, C.L.E. Dynamic Modeling of an Industrial Electric Arc Furnace. Industrial & Engineering Chemistry Research 2005, 44, 8067–8083, doi:10.1021/ie050101b.

[6] Arnout, S.; Verhaeghe, F.; Blanpain, B.; Wollants, P.; Hendrickx, R.; Heylen, G. A Thermodynamic Model of the EAF Process for Stainless Steel. steel research int. 2006, 77, 317–323, doi:10.1002/srin.200606392.

[7] Kho, T.S.; Swinbourne, D.R.; Blanpain, B.; Arnout, S.; and Langberg, D. Understanding stainless steelmaking through computational thermodynamics Part 1: electric arc furnace melting. Mineral Processing and Extractive Metallurgy 2010, 119, 1–8, doi:10.1179/174328509X431454.

[8] Logar, V.; Dovžan, D.; Škrjanc, I. Modeling and Validation of an Electric Arc Furnace: Part 1, Heat and Mass Transfer. ISIJ International 2012, 52, 402–412, doi:10.2355/isijinternational.52.402.

[9] Logar, V.; Dovžan, D.; Škrjanc, I. Modeling and Validation of an Electric Arc Furnace: Part 2, Thermo-chemistry. ISIJ International 2012, 52, 413–423, doi:10.2355/isijinternational.52.413.

[10] Meier, T. Modellierung und Simulation des Elektrolichtbogenofens, 2016, ISBN 978-3-95886-129-9.

[11] Hay, T. Mathematische Modellierung des Elektrostahlverfahrens. Dissertation; RWTH Aachen, Aachen, 2021.

[12] Hay, T.; Reimann, A.; Echterhof, T. Improving the Modeling of Slag and Steel Bath Chemistry in an Electric Arc Furnace Process Model. Metallurgical and Materials Transactions B 2019, 50, 2377–2388, doi:10.1007/s11663-019-01632-x.

[13] Linde. PSA: Pressure Swing Adsorption Plants. Available online: https://assets.linde.com/-/media/global/engineering/engineering/home/products-and-services/process-plants/ha-_1_1_e_12_150dpi_nb.pdf (accessed on 16 April 2025).

[14] Elsabagh, S. Analyse und Optimierung der Schaumschlackenfahrweise eines Gleichstrom-Elektrolichtbogenofens mittels Machine Learning Verfahren, 2022, ISBN 978-3-96463-026-1.

[15] Kühn, R. Untersuchungen zum Energieumsatz in einem Gleichstromlichtbogenofen zur Stahlerzeugung; Shaker Verlag: Düren, 2003, ISBN 978-3-8322-2197-3.

[16] Fruehan, R.J. Fundamentals and Practice for Producing Low Nitrogen Steels. ISIJ International 1996, 36, S58-S61, doi:10.2355/isijinternational.36.Suppl_S58.

[17] Pal, J. Thermodynamic analysis of nitrogen removal in EAF by DRI fines injection. Ironmaking & Steelmaking 2006, 33, 465–470, doi:10.1179/174328106X149824.

[18] Neural Networks and Deep Learning: A Textbook; Aggarwal, C.C., Ed.; Springer International Publishing: Cham, 2023, ISBN 978-3-031-29642-0.

[19] Szandała, T. Review and Comparison of Commonly Used Activation Functions for Deep Neural Networks. In Bio-inspired Neurocomputing; Bhoi, Akash Kumar and Mallick, Pradeep Kumar and Liu, Chuan-Ming and Balas, Valentina E., Ed.; Springer Singapore: Singapore, 2021; pp 203–224, ISBN 978-981-15-5495-7.

[20] Goodfellow, I.; Bengio Yoshua; Courville, A. Deep Learning; MIT Press, 2016, ISBN 9780262035613.

[21] Bowman, B.; Krüger, K. Arc Furnace Physics, 1st ed.; Stahleisen, 2009, ISBN 978-3514007680.

# Iron recovery from Waelz slag through biogenic carbothermic reduction

**Gianluca Dall'Osto[1*], Davide Mombelli[1], Sara Scolari[1] and Carlo Mapelli[1]**

[1] Dipartimento di Meccanica, Politecnico di Milano, Milano, Italy

*E-mail: Gianluca.dallosto@polimi.it

**Abstract.** Waelz slag (WS) can be considered as one of the iron and steel related residues that have not yet been fully valorised. With more than 600-800 kg of WS generated per ton of electric arc furnace dust processed in the Waelz process, and a residual iron content of 10-40 wt.%, the iron recovery from WS by carbothermic reduction can represent an intriguing valorisation pathway. In this study, WS was mixed with an olive pomace derived biochar in three recipes with increasing $C/(Fe_2O_3+ZnO)$ ratios and subsequently agglomerated. The reduction behaviour (mass loss, swelling, degree of metallization, mineralogy and chemistry) of the carbon composite agglomerates was evaluated at 1000, 1200 and 1400 °C under inert atmosphere and compared with the corresponding performance of coke as a reducing agent. The results highlighted the possibility of iron recovery from WS at 1400 °C and the comparable reduction behaviour in the case of using either a biogenic or fossil carbon-based reducing agent. The optimal $C/(Fe_2O_3+ZnO)$ ratio was found to be 0.262, with a degree of metallization of 91 wt.% achieved at 1400 °C using biochar. The mineralogy of the slag produced, mainly composed of calcium aluminates and calcium silicates, showed high desulphurization properties, thus protecting the recovered iron particles from pollution. Nevertheless, the absence of slag-metal separation at each thermal treatment temperature suggests the need for pretreatment of the WS or the addition of fluxes to the agglomerate recipe.

## 1. Introduction

Waelz slag (WS) can be considered as one of the most critical residues closely associated with either the zinc recovery cycle or the iron and steel industry (ISI). It is generated during the pyrometallurgical extraction of zinc from electric arc furnace dust (EAFD) using the Waelz process. Specifically, about 600-800 kg of WS is generated per ton of EAFD treated, which in turn is generated at a rate of 15-25 kg per ton of steel produced by EAF [1]. Considering that about 5-8 Mton of EAFD are produced annually worldwide, and that this amount is expected to reach at least 18 Mton by 2050 due to the increasing share of EAF, about 10.8-14.4 Mton of WS will have to be properly managed each year from 2050 onwards. In the European context, it was estimated that about 0.5 Mton of WS were produced in 2000 [2,3].

To date, although WS is mostly landfilled and thus not valorised, it has started to gain some attraction to be used as inert or filler material in the construction industry, historically one of the major acceptors of ISI-related slag [4]. However, the main limitation of such possibility is related to the national thresholds of leachable heavy elements (e.g. As, Cd, $Cr^{6+}$, Cu, Pb, Ba and V) [5]. For example, Quijorna et al. [6] pointed out that to comply with the Spanish regulatory leaching limits for Mo, it is possible to add a maximum of 30 wt.% of WS in red clay bricks. Vegas et al. [7] studied several possibilities of using WS in road construction, highlighting that the critical property for its use as a sub-base in pavement structural layers is the sulphur content, while Antrekowitsch et al. [8] argued that in case of high chromium and nickel content WS, such practice would be unsuitable due to the current strict legal regulations. Finally, Khanapur et al. [9] suggested that the substitution of 50 wt.% of the fine recycled concrete aggregates for the concrete production with WS can improve the mechanical and durability of the concrete itself, as well as fully comply with the leaching limits allowed by the United States code of federal regulations.

Besides the aforementioned valorisation pathways of the WS, studies have started to investigate its direct recycling into the ISI with a specific attention of its use as a secondary raw material for EAFs [10]. Indeed, WS is essentially an iron-rich slag with a percentage of iron oxides ranging from 10 to 40 wt.%, depending on the EAFD mixture used, their composition, and the operating conditions of the Waelz process itself. It can therefore be estimated that potentially more than 200 kton of secondary iron per year could be effectively returned to the steelmaking cycle. Furthermore, on the one hand, the use of sand as a binder in the feed will result in a basicity index (BI), defined as $(CaO+MgO)/SiO_2$, lower than one, and thus to an acidic slag. On the other, the use of lime will result in a BI greater than one and therefore a basic slag. Finally, if the zinc extraction yield is low, the corresponding oxide also ends up in the WS, which could hinder to some degree its charging into metallurgical furnaces [3].

On this regard, the calculations carried out by Fröhling et al. [11] highlighted that the direct reintroduction of WS in EAFs would allow a reduction in slag forming fluxes of 25-30%, accompanied by an increase in the quality of the EAFD, at the expense of an increase in carbon and electricity demand per ton of steel of 75% and 10-15%, respectively. However, no information about the chemistry of the WS were provided in the study. Concurrently the iron recovery and zinc removal from WS by means of gaseous flows have been investigated at several levels [12–14]. Specifically, although a zinc extraction yield of up to 90% and maximum metallization degree of 95 wt.% of the reduced WS were achieved when hydrogen was used as a reducing agent, the investigation was stopped at the pilot scale. This was due to the high costs associated with the reducing gas reforming unit and the presence of elements harmful to metallurgical furnaces (e.g. Zn, Pb) in the reduced WS, which restricted it to dumping [14]. Finally, the recovery of iron from a WS rich in lime and coke breeze by carbothermic reduction smelting was successfully carried out by Grudinsky et al. [15]. Specifically, the WS used was characterized by Ca, C, Fe and Zn contents of 9.93, 18.7, 26.4 and 0.94 wt.%, respectively. Nevertheless, the process required a temperature of 1650 °C for 30 min and the addition of acidic fluxes (18 wt.% of $SiO_2$) to achieve a proper slag-metal separation.

Pursuing the feasibility of iron recovery from critical industrial residues (e.g. red mud [16,17], mill scale [18,19], jarosite [20,21], and WS itself [15]) by means of carbothermic reduction, this work explores the metallurgical behaviour of carbon composite agglomerate (CCA), made exclusively from WS and a solid carbonaceous source, for the effective secondary iron production. In order to fill the gaps currently existing in the literature, special attention has been paid to the optimal $C/(Fe_2O_3+ZnO)$ ratio of the CCA and to the reduction efficiency of biogenic carbon sources with respect to that of fossil carbon sources. The choice of using non-fossil carbon sources was made to further develop the knowledge on sustainable alternative iron production processes, as well as to strengthen the industrial symbiosis between ISI and major biomass producing sectors (e.g. wood/furniture, agriculture and wastewater treatment plants).

Journal of Physics: Conference Series **3050** (2025) 012003 doi:10.1088/1742-6596/3050/1/012003

## 2. Material and methods

*2.1 Waelz slag characterization*

The iron-bearing matrix used in this study, Waelz slag, was collected after cooling in the open air at the end of the Waelz process itself. Prior to characterization, the WS was homogenized by grinding in a planetary ball mill (Retsch PM 400, Retsch, Technology GmbH, Haan, Germany) with zirconia balls and subsequently sieved to very fine sand (63–125 μm), the latter to optimize the carbothermic reduction kinetics [20].

A wavelength dispersive X-ray fluorescence (WD-XRF) spectrometer (Bruker S8 TIGER, Bruker AXS GmbH, Karlsruhe, Germany) was used for elemental analysis, while an X-ray diffraction (XRD) analyser (Rigaku SmartLab SE diffractometer, Rigaku Corporation, Tokyo, Japan) was used to obtain the mineralogical composition of the studied WS. Specifically, the diffraction pattern was acquired between 15 and 90° 2θ, with a scan rate of 2° $min^{-1}$, step size of 0.02°, sample rotation of 60 rpm and using a Cu-Kα radiation (λ=1.54 Å) source with a 1D XRF suppression detector (D/Tex 250), whereas peak identification was performed using the Crystallography Open Database (COD). The visual appearance, elemental composition and mineralogy of the WS are shown in **Figure 1.**

(a)

| Element | wt.% |
|---|---|
| $Al_2O_3$ | 3.75 |
| CaO | 20.7 |
| $Cr_2O_3$ | 0.95 |
| CuO | 0.32 |
| $Fe_2O_3$[a] | 39.40 |
| MgO | 2.54 |
| MnO | 3.18 |
| $SO_3$ | 4.86 |
| $SiO_2$ | 7.04 |
| ZnO | 12.5 |
| Alkali | 2.44 |
| LOI @ 600 °C | -1.18 |
| LOI @ 1050 °C | -2.30 |

[a]Total iron (Fe+FeO) expressed as $Fe_2O_3$.

(b)

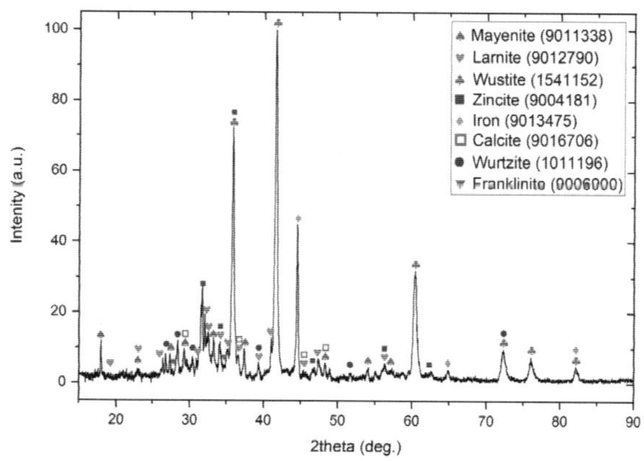


(c)

**Figure 1.** Summary of Waelz slag characterization (a) visual appearance, (b) elemental composition by X-ray fluorescence (WD-XRF) and (c) mineralogical composition by diffraction pattern.

The elemental analysis (**Figure 1.a**) revealed iron as the major constituent, followed by calcium and zinc, due to the addition of lime during the Waelz process and incomplete recovery of the zincite contained in the EAFD, respectively. A significant concentration of sulphur was also observed, which can be attributed to the addition of coke during the Waelz process. Finally, the presence of small amounts of chlorine, fluorine, sodium and potassium was attributed to the absence of any washing stage. The comparison with the XRD pattern (**Figure 1.b**) showed that iron was mainly present as wustite (FeO) and to a lesser extent as metallic iron (Fe) or franklinite ($Fe_2ZnO_4$), while zinc was present as zincite (ZnO) or bound with sulphur as wurtzite (ZnS). On the other hand, calcium was present together with either aluminium or silicon as mayenite ($Ca_{12}Al_{14}O_{33}$) and larnite ($Ca_2O_4Si$), respectively.

*2.2 Recipe preparation and carbon composite agglomerate manufacturing*
Two carbon matrices were used as reducing agents for the oxides contained in the WS: a biochar obtained from the pyrolysis at 750 °C of olive pomace (69.9 wt.%C) and a metallurgical coke (87.5 wt.%C). Specifically, their characterization was discussed in a previous work [22].

**Table 1.** Labelling and composition of carbon composite agglomerate (CCA) recipes investigated.

| Carbonaceous matrix | Olive pomace biochar | | | Metallurgical coke | | |
|---|---|---|---|---|---|---|
| CCA recipe ID | B1 | B2 | B3 | C1 | C2 | C3 |
| Reducing agent in CCA, wt.% | 15.77 | 27.20 | 37.50 | 11.44 | 20.59 | 29.29 |
| Waelz slag in CCA, wt.% | 84.23 | 72.80 | 62.50 | 88.56 | 79.50 | 70.71 |
| $C/(Fe_2O_3+ZnO)$ | 0.262 | 0.523 | 0.840 | 0.262 | 0.523 | 0.840 |

Due to the concomitant reduction of iron and zinc at temperatures above 1000 °C and the significant presence of the latter in the slag studied, each reducing agent was added to the WS to obtain three recipes characterized by a $C/(Fe_2O_3+ZnO)$ ratio of 0.262, 0.523 and 0.840, respectively. An organic binder made of gelatinized corn starch was added to the powders in a dosage corresponding to 5% of the total mass. After mechanical homogenization, the mixture was compacted into a CCA of briquette morphology (20x20 mm) at 40 MPa with a compaction rate of 20 mm $min^{-1}$ and dwell time of 2 min using a modified uniaxial tensile machine (MTS Exceed Series 40, MTS Systems, Eden Prairie, Minnesota). Finally, the green CCA were cured for 20 days to ensure the expulsion of the residual water contained in the binder and the stabilization of the mass. The labelling and characteristics of each CCA are shown in **Table 1**.

*2.3 Reduction behavior*
The reduction behavior of the CCAs was evaluated according to their (i) mass loss, (ii) swelling (iii) metallization degree evolution between 1000 and 1400 °C. Lastly, the sample mineralogy and chemistry at 1400 °C was investigated.

One sample at a time was tested using the crucible setup previously described in [19]. Each thermal treatment was performed in a lift-bottom furnace (Nabertherm LHT 02/17, Nabertherm, Lilienthal, Germany) with a heating rate of 100 °C $min^{-1}$, a dwell time of 30 min at the target temperature and a natural cooling of the furnace. An argon flow of 10 Nl $h^{-1}$ was blown throughout the treatment to provide an inert atmosphere.

*2.3.1 Mass loss evaluation*
The mass loss was preferred as a behavioral parameter due to the bias that the biogenic carbon gasification (e.g., expulsion of volatile matter) would bring to the reduction degree and evaluated according to **Equation 1**.

$$mass\ loss\ (wt.\%) = \left(1 - \frac{w_i}{w_o - w_b}\right) \times 100 \qquad \textbf{Equation 1}$$

where $w_i$ is the mass of the sample after the thermal cycle, $w_0$ is the initial mass of the sample and $w_b$ is the mass of the binder used in the sample.

*2.3.2 Swelling behaviour*

Before and after thermal treatment, each sample was spatially reconstructed in 3D by means of photogrammetry (Agisoft Metashape software version 1.8.5, Agisoft LLC) to evaluate the corresponding apparent volume, assuming the absence of internal porosity. Consequently, the swelling behavior was evaluated according to **Equation 2**.

$$swelling\ (vol.\%) = \left(\frac{V_i}{V_o} - 1\right) \times 100 \qquad \textbf{Equation 2}$$

where $V_i$ is the volume of the sample after the thermal cycle, $V_0$ is the initial volume of the sample.

*2.3.3 Degree of metallization*

The degree of metallization (DoM), defined as the ratio of metallic iron over the total iron, was evaluated by means quantitative metallurgy. Five scanning electron microscopy micrographs in backscattered electron (SEM-BSE) configuration were taken stochastically and post-processed by means of ImageJ software (version 1.53t). **Figure 2** shows an example of the methodology followed.

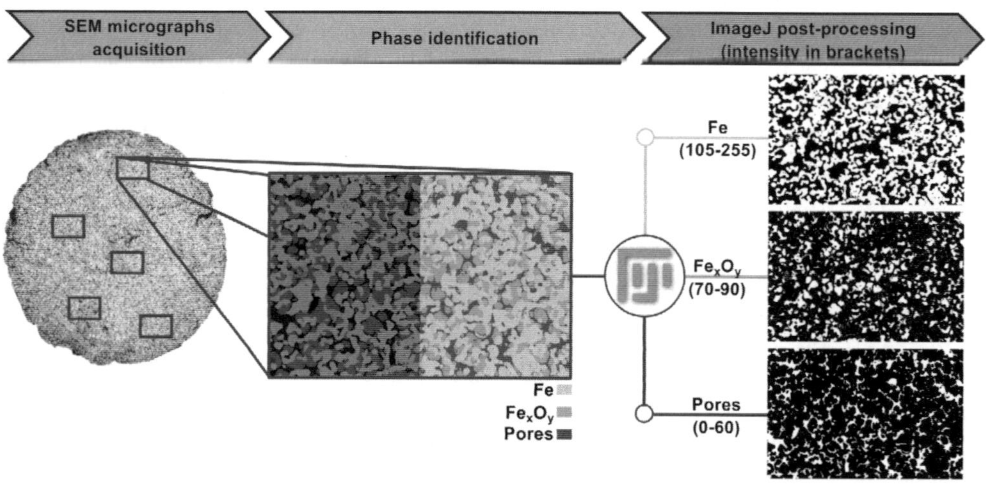


**Figure 2.** Example of the methodology followed for the evaluation of the degree of metallization.

The intensity threshold of each phase was evaluated and applied to the corresponding micrograph to calculate the number of pixels occupied by the specific phase. Consequently, the DoM was calculated according to **Equation 3**.

$$DoM\ (\%) = \frac{Fe_M}{Fe_T} \times 100 = \frac{n_{Fe} \cdot \rho_{Fe}}{n_{Fe} \cdot \rho_{Fe} + n_{FeO} \cdot \rho_{FeO} \cdot \chi} \times 100 \qquad \textbf{Equation 3}$$

Journal of Physics: Conference Series **3050** (2025) 012003
doi:10.1088/1742-6596/3050/1/012003

where $Fe_M$ is the metallic iron, $Fe_T$ is the total iron, $n_{Fe}$ is the number of pixels attributed to the metallic iron, $\rho_{Fe}$ is the metallic iron density, $n_{FeO}$ is the number of pixels attributed to the oxide iron, $\rho_{FeO}$ is the oxide iron density and $\chi$ is the weight percentage of iron in the specific oxide.

*2.3.4 Mineralogical and phase distribution evaluation*

The mineralogy of the reduced CCA at 1400 °C was evaluated through XRD analysis. If feasible, the sample was ground to a particle size of less than 100 μm or otherwise cut in half to expose its cross-section. The diffraction spectrum was acquired from 15 to 90° 2θ at 1° $min^{-1}$ with a step size of 0.02° and a rotation of 120 rpm to reduce the texture contribution behavior.

Scanning electron microscopy coupled with energy dispersive X-ray spectroscopy (SEM-EDS) was performed using a Zeiss Sigma 300 FEG-SEM (Carl Zeiss AG, Jena, Germany) to map the phase distribution of the sample (e.g. metallic iron, iron oxide, slag) and to evaluate the general chemical composition.

## 3. Results

*3.1 Macroscopic appearance of the reduced agglomerates*

The visual appearance of the CCAs after thermal treatments is shown in **Figure 3**. To better highlight the presence of cracks, images are provided in grayscale [23].

**Figure 3.** Visual appearance of the carbon composite agglomerates at 1000, 1200 and 1400 °C under inert atmosphere.

Although they were able to maintain a mechanical stability and low friability, the gases generated during the oxide reduction contained in the WS (e.g. $FeO_{(s)}+C_{(s)}\rightarrow Fe_{(s)}+CO_{(g)}$, $ZnO_{(s)}+C_{(s)}\rightarrow Zn_{(g)}+CO_{(g)}$) were responsible for the expansion of the internal porosities, while their expulsion from the CCA matrix for the creation of cracks that characterized the reduced briquette surface [24]. Similar to the results obtained by Grudinsky et al [15], no clear separation between the recovered metal and the slag was obtained even at 1400 °C due to the high viscosity of the slag formed and the possible presence of unreacted carbon in it.

*3.2 Mass loss*

The overview of the mass losses of the CCAs at 1000, 1200 and 1400 °C under inert atmosphere is provided in **Figure 4**.

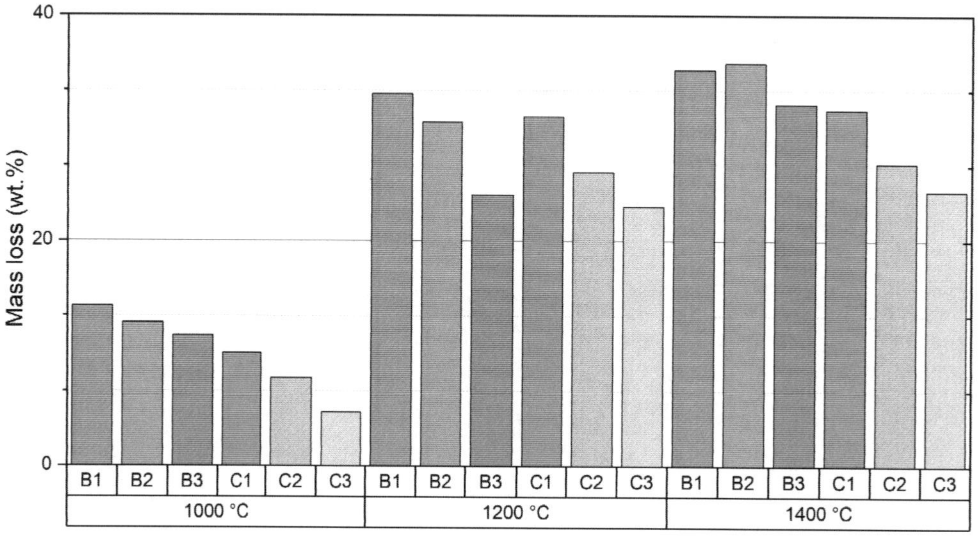


**Figure 4.** Mass loss of the carbon composite agglomerates at 1000, 1200 and 1400 °C.

As a rule of thumb, the entire set of B recipes was characterized by a higher mass loss than the C recipes. This was attributed to both the higher content of reducing agent present in the recipe (**Table 1**) and amount of volatile matter present in the olive pomace biochar compared to metallurgical coke (51.90 vs. 5.20 wt.%) [18,22]. Still in this regard, the volatile matter was responsible of the higher presence of cracks in the B recipe samples (**Figure 3**).

As expected from the reduction kinetics of CCAs, the mass loss trend of each formulation was found to be related to the thermal treatment temperature [25,26]. From the initial value at 1000 °C, the mass loss is subjected to a steep increase at 1200 °C and subsequent stabilization at 1400 °C. The whole set of B recipes was characterized by a twofold increase in mass loss from 1000 to 1200 °C (from 12-14 to 32-35 wt.%), while the C formulations showed a threefold increase (from 5-10 to 24-32 wt.%). The higher activation energy of metallurgical coke (116.18 kJ $mol^{-1}$) compared to that of biochar (86.76 kJ $mol^{-1}$) together with the previously addressed volatile matter content were most probably responsible for the difference in relative mass loss increase between 1000 and 1200 °C [22,27].

Notably, an inverse proportional trend was observed between the mass loss and the $C/(Fe_2O_3+ZnO)$ ratio. This peculiar behaviour was attributed to the excess of carbon present in the recipes having a $C/(Fe_2O_3+ZnO)$ ratio of 0.523 and 0.840, respectively. In other words, after all the removable oxygen in the WS is consumed, the carbon is not able to escape from the CCA matrix as carbon monoxide and remains trapped inside the latter [28]. Nevertheless, its presence can be essential for the following carburization of the reduced iron particles [29].

### *3.3 Swelling behaviour*

The above assumption is confirmed by the swelling trend as a function of the C/($Fe_2O_3$+ZnO) ratio and the thermal treatment temperature, the latter shown in **Figure 5**.

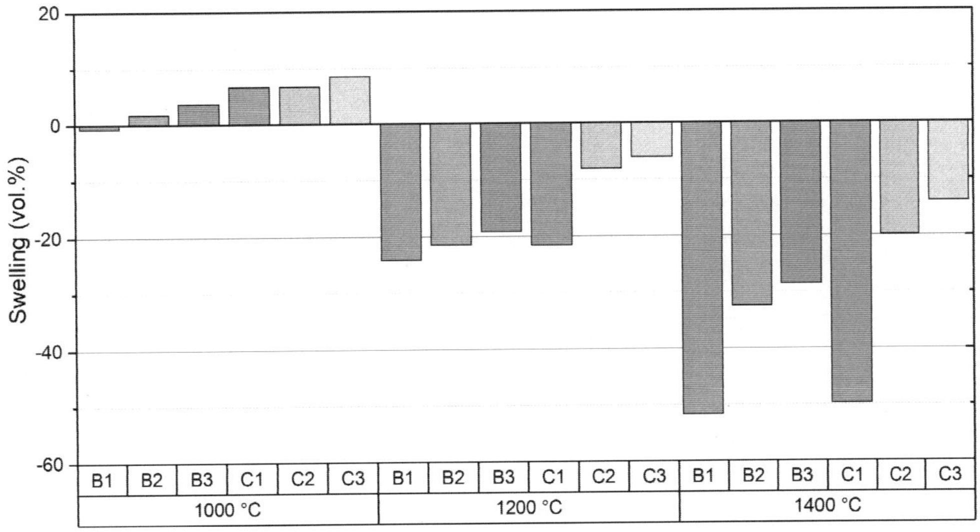


**Figure 5.** Swelling behaviour of the carbon composite agglomerates at 1000, 1200 and 1400 °C.

Indeed, although all the recipes were subjected to a shrinkage proportional to the thermal treatment temperature, those with a C/($Fe_2O_3$+ZnO) ratio of 0.262 were characterized by the highest volume reduction, followed by those with a ratio of 0.523 and 0.840. This was due to the presence of the trapped unreacted carbon in the B2, C2, B3 and C3 recipes, which act as a filler [29]. Furthermore, the higher gasification tendency of the olive pomace biochar led the samples belonging to the B recipes to a higher volume reduction. Specifically, the maximum swelling was observed at 1000 °C for the C3 sample (8.57 vol.%), which is fully in line with the normal swelling of CCA when in metallurgical furnaces (10-20 vol.%) [30]. The highest volume reductions at 1400 °C were achieved by the B1 and C1 samples with - 51.69 and - 49.73 vol.%, respectively.

### *3.4 Degree of metallization and cross-section observation*

The degree of metallization achieved by the CCA at 1000, 1200 and 1400 °C under inert atmosphere is given in **Figure 6**.

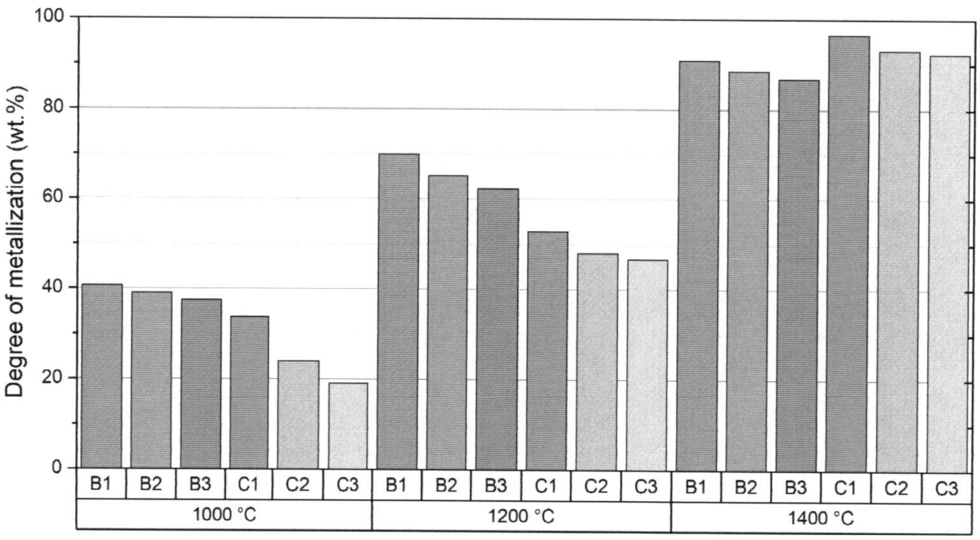


**Figure 6.** Degree of metallization of the carbon composite agglomerates at 1000, 1200 and 1400 °C.

According to the general theory of carbothermic reduction, the predominant reduction mechanism up to about 1000-1100 °C corresponds to the gas-solid reduction ($MeO_{(s)}+CO_{(g)}\rightarrow Me_{(s)}+CO_{2(g)}$), which is directly related to the amount of volatile matter and inversely related to the gasification tendency of the carbonaceous matrix used. On the contrary, the final stages are controlled by the solid-solid reduction mechanism ($2MeO_{(s)}+C_{(s)}\rightarrow 2Me_{(s)}+CO_{2(g)}$), which is directly related to the amount of fixed carbon present in the reducing agent [28,31]. The carbon dioxide produced can then react with the solid carbon to produce more carbon monoxide for gas-solid reduction ($C_{(s)}+CO_{2(g)}\rightarrow 2CO_{(g)}$) or for indirect carburization of the reduced iron particles ($3Fe_{(s)}+2CO_{(g)}=Fe_3C_{(s)}+CO_{2(g)}$). Direct carburization of the iron particles can also occur in the presence of excess free solid carbon [32].

Thus, olive pomace biochar, characterized by higher volatile matter and lower gasification activation energy, was advantageous in terms of reduction efficiency with respect to coke at both 1000 and 1200 °C, while the opposite occurred at 1400 °C, explaining the observed DoM trend.

Consistent with the work of Misha et al. [33] on iron ore based CCA reduction in rotary heart furnaces and Mombelli et al. [34] on carbothermic reduction of jarosite waste by blast furnace sludges, the higher the $C/(Fe_2O_3+ZnO)$ ratio, the lower the DoM achieved. According to their results, although an excess of carbon shifts the iron reduction temperature toward lower values, an under stoichiometric-to-stoichiometric ratio avoids excessive carbon gasification in the early stages of carbothermic reduction while promoting the efficiency of the solid-solid reduction mechanism and, in turn, gaseous-solid reduction at temperatures above 1200 °C.

Journal of Physics: Conference Series **3050** (2025) 012003 doi:10.1088/1742-6596/3050/1/012003

The above assumptions appear to be confirmed by the micrographs of the reduced CCA cross-section shown in **Figure 7**.

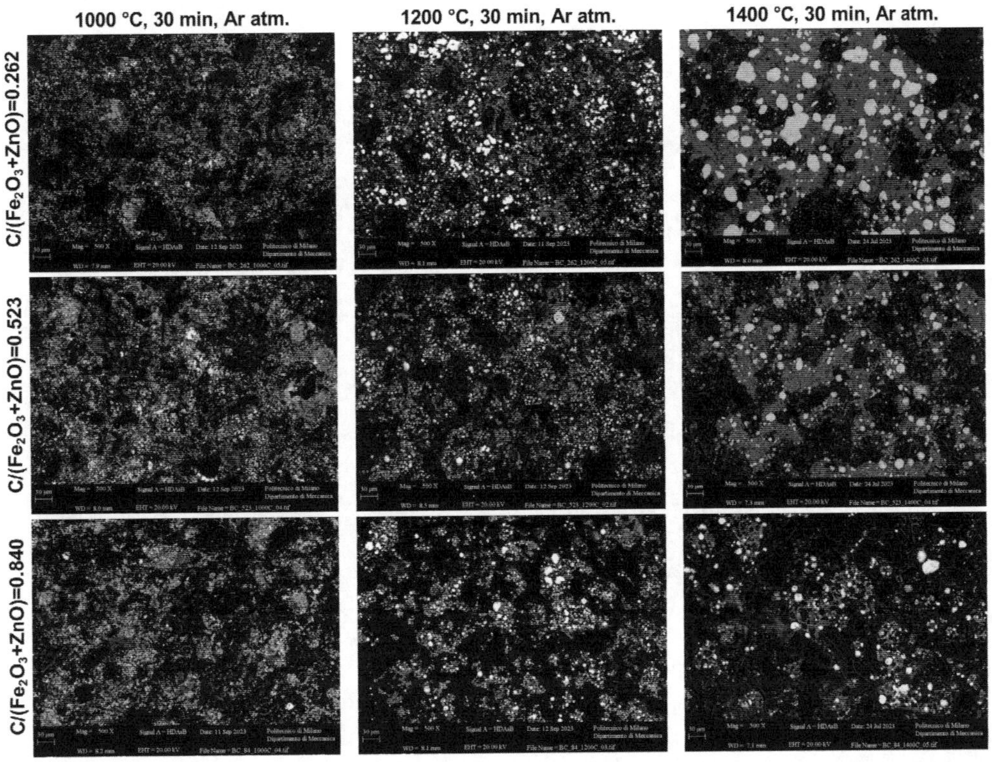


**Figure 7.** SEM-BSE micrographs of B recipe carbon composite agglomerates as a function of thermal treatment temperature and C/($Fe_2O_3$+ZnO) ratio (light grey particles: reduced iron oxides, grey particles: unreduced oxides, dark grey and black areas: residual carbon and porosities).

On the one hand, no major differences were observed at 1000 °C as a function of the C/($Fe_2O_3$+ZnO) ratio. Each recipe was characterized by the presence of the first reduced iron particles and porosities due to gas-solid reduction. On the other, the first behavioural heterogeneities appeared at 1200 °C. As the C/($Fe_2O_3$+ZnO) ratio increased, the CCA matrix appeared more porous due to the increased gasification rate caused by the excess of carbon, and the reduced iron particles appeared smaller due to the yet unreacted carbon, which hindered the sintering of the particles acting as a filler [29]. These differences were even more evident from the observation of the corresponding CCAs cross-section at 1400 °C. Specifically, average reduced iron particle diameters of 66, 34, and 19 μm and areal porosities of 21.5, 28.1, and 45.8%were observed for reduced C1, C2, and C3 samples, respectively.

Journal of Physics: Conference Series **3050** (2025) 012003 doi:10.1088/1742-6596/3050/1/012003

*3.5 Mineralogical and phase distribution characterization of the reduced products*

The XRD pattern of the CCA recipes at 1400 °C is shown in **Figure 8**.

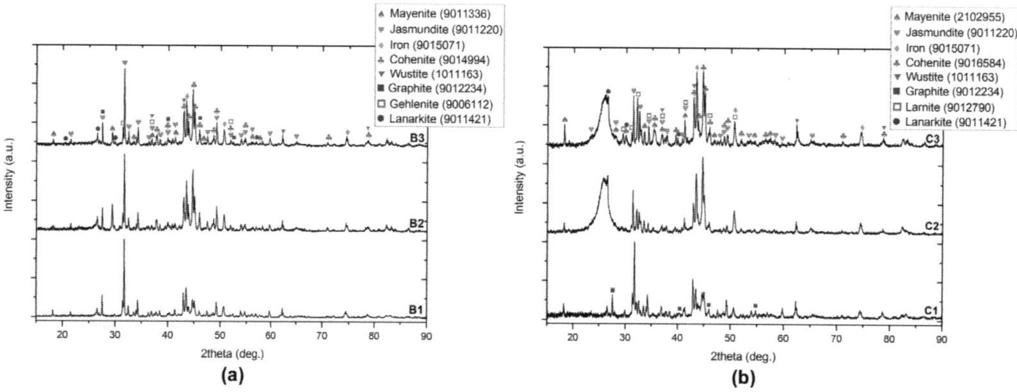


**Figure 8.** Diffraction pattern of the carbon composite agglomerates at 1400 °C (DB card number in parenthesis).

Overall, each recipe was characterized by a comparable mineralogy. Iron was present as either metallic iron (Fe), cohenite ($Fe_3C$), or, to a lesser extent, wustite (FeO). Carbon was present either as graphite or as amorphous carbon. The major constituents identified in the reduced CCA slag were based on Ca-, Si- and Al- compounds. On the one hand, mayenite ($12CaO{\cdot}7Al_2O_3$), and jasmundite ($Ca_{11}(SiO_4)_4O_2S$) were present in all formulations. On the other hand, larnite ($Ca_2SiO_4$) was identified only in the sample belonging to the C recipe, while gehlenite ($Ca_2Al_2SiO_7$) in those belonging to the B recipe. Finally, the minor presence of lead sulphate in the form of lanarkite ($Pb_2(SO_4)O$) was also observed in all sample slag.

The comparison of the relative peak intensities of metallic iron and cohenite revealed an increasing carburization of the reduced CCAs as a function of the C/($Fe_2O_3$+ZnO) ratio. Interestingly, the B1 and B2 samples showed a higher carburization at 1400 °C than their counterparts belonging to the C recipe. This behaviour was attributed to the faster reduction and metallization kinetics commonly observed when using a biochar reductant rather than metallurgical coke [35]. Indeed, the higher DoM observed at 1200 °C for the B recipe samples allowed an earlier initiation of the direct carburization, driven by the unreacted carbon present in the CCA, as well as the gaseous carburization, driven by the carbon monoxide generated during the carbothermic reduction of the oxides still present [32].

The increasing presence of residual carbon as a function of the C/($Fe_2O_3$+ZnO) ratio was confirmed by the corresponding increase in intensity of the carbonaceous phases identified in the diffraction patterns. On the one hand, the samples belonging to the B recipes, as well as the C1 sample, were characterized exclusively by the presence of graphite with its characteristic peak at 26.7°. On the other, the hump in the 22-28° range, observed in the C2 and C3 samples, indicated the presence of amorphous carbon. This internal difference within the C recipe samples was attributed to the graphitization process that can occur in the temperature range used in this work for the CCA reduction, and to the lower amount of carbon present in the C1 recipe itself compared to the C2 and C3 recipes (**Table 1**) [36].

Finally, the slag mineralogy was consistent with that typically observed in highly sulphurised steel slag and with the initial WS mineralogy (**Figure 1.c**), with mayenite present both before and after CCA reduction [37]. On the contrary, and similar to what occurs during clinker production in cement kilns when high amounts of sulphur are present, the solid-state reaction at temperatures above 1000 °C between the larnite and the sulphur released by wurtzite (ZnS) decomposition and oxidation has led to the synthesis of jasmundite [38,39].

**Figure 9** shows the elemental and overlayed SEM-EDS maps of the B1 sample cross-section taken as the most representative.

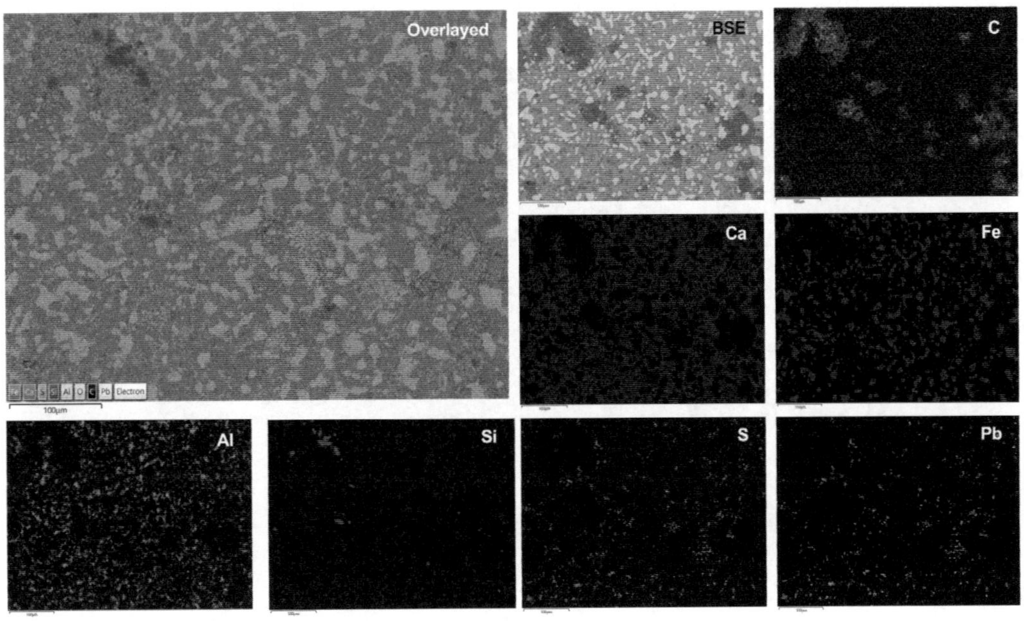


**Figure 9.** SEM-EDS mapping of the C1 sample cross-section.

The analysis revealed the presence of metallic iron and carburized iron particles embedded in the slag matrix, mainly composed of jasmundite and mayenite, with the presence of either unreduced iron, unreacted carbon and lead sulphate. Finally, no zinc was observed, confirming the assumption of its complete reduction and evaporation during the carbothermic reduction.

## 4. Discussion, process feasibility and optimization

The study has confirmed the feasibility of iron recovery from WS through carbothermic reduction process, highlighting also the associated criticalities. The WS was agglomerated into CCAs with increasing C/($Fe_2O_3$+ZnO) ratio using two different reducing agent matrices: a biochar obtained from olive pomace pyrolysis at 750 °C and a metallurgical grade coke. The CCAs were then thermally treated at 1000, 1200 and 1400 °C to understand their reduction behaviour and the most appropriate C/($Fe_2O_3$+ZnO) ratio to maximize iron recovery while minimizing the amount of reducing agent used.

Increasing the thermal treatment temperature and decreasing the C/($Fe_2O_3$+ZnO) ratio from 0.860 to 0.262 improved the degree of metallization and minimized the amount of unreacted carbon. Specifically, when biochar was used as reducing agent, the values at 1400 °C were slightly lower than those commonly observed for commercial direct reduced iron (DRI), 87-91 wt.% versus 92-96 wt.%, respectively. [40] On the other hand, those obtained with coke were fully comparable to the commercial values.

The significant percentage of non-ferrous elements (e.g. Ca, Zn, Si, Al, S) present in the starting WS has led to the formation in each recipe of a calcium aluminate and calcium silicate-based slag. Although the latter was able to protect the reduced iron from sulfidation by forming jasmundite, it constituted a large proportion of the reduced CCA and its chemistry did not allow proper separation from the metallic fraction. Finally, no trace of zinc, either metallic or bounded, was observed, highlighting its complete removal during the carbothermic reduction.

Overall, among the recipes, the C/($Fe_2O_3$+ZnO) ratio of 0.262 can be considered the most suitable for the recovery of iron contained in WS. While, between the matrices, olive pomace biochar, although characterized by a lower metallization capacity than coke at 1400 °C, would avoid the environmental impact associated with the use of the latter. In addition, the use of olive pomace biochar would mean the establishment of an industrial link with the agricultural sector.

Developing this hypothesis in the Italian scenario, given the importance of the olive oil industry in the national economy (20% of the world production is located in Italy [41]), the major production of steel by EAF (85.8% of the Italian ISI, 18.1 Mton per year [42]) and to the presence of a Waelz process plant, the estimated secondary iron production from WS would be 80 kton per year, hence able to cover 4% of the annual import of clean iron sources (2 Mt) required by the EAF. The calculation was based on the investigated WS chemistry, a WS production of 700 kg per ton of EAFD processes and an EAFD of 20 kg per ton of steel produced by EAFD. This would require about 48 kton of olive pomace biochar per year.

Still in an optics of industrial applicability, the direct charging of the reduced CCAs as secondary iron source in EAF should be properly evaluated. Among the criticalities and process optimizations to be addressed, the first one relates to the achievement of a proper metal-slag separation during the carbothermic reduction. Indeed, the high amount of slag observed with reduced CCA would have a negative impact on most of the process parameters by reducing yield, increasing energy consumption and adding flux once charged [40]. Secondly, since in the carbothermic reduction process the entire amount of zinc present in the starting WS was evaporated and not recovered, this means the loss of one of the most important raw materials for the galvanizing, die casting and brass industries [43].

Therefore, among the possible process optimization paths to be followed, the pre-processing of WS by means of gravimetric separation, magnetic separation or hydrometallurgy could represent a cost-effective solution for the upgrading of the ferrous fraction as well as the selective recovery of the zinc the other valuable elements contained in the WS, as is currently being investigated for EAFD [44,45]. An opposite possibility would be the flux addition to the CCA recipe to lower the melting temperature of the slag and thus achieve an effective slag-metal separation [15]. However, in this scenario, a flue gas capture system should be designed for the subsequent zinc oxide dust condensation and recovery, similar to what is already done in the Waelz process.

## 5. Conclusions

Based on the reduction behaviour at 1000, 1200 and 1400 °C of several carbon composite agglomerate recipes using either olive pomace biochar or metallurgical coke as reducing agents for the iron recovery from Waelz slag, the following conclusions can be drawn:

- At each thermal treatment temperature, the reduction behaviour was comparable among the recipes using olive pomace biochar or coke as reducing agent.
- Carbothermic reduction at 1400°C is capable of effectively reducing both Fe- and Zn-bearing oxides/compounds (e.g. FeO, ZnO and ZnS) present in Waelz slag, regardless of the type and amount of reducing agent introduced in the formulation.
- The recipes tailored to a C/($Fe_2O_3$+ZnO) ratio equal to 0.262 achieved the highest degree of metallization at 1400 °C, 91 and 96 wt.%, namely using olive pomace and metallurgical coke as reducing agents respectively.
- Increasing the C/($Fe_2O_3$+ZnO) ratio to 0.523 and 0.840 resulted in a degree of metallization decrease and the presence of unreacted carbon embedded in the reduced product matrix, which inhibited both the shrinkage and the coalescence of the recovered iron particles.
- Although a proper slag-metal separation phase did not occur, the slag generated at 1400 °C showed a desulphurisation capacity able to protected the reduced iron particles from pollution.
- Due to the high amount of slag present in the reduced product either the pre-processing of the Waelz slag to upgrade the ferrous fraction and selectively recover part of the non-ferrous elements, or the addition of fluxes in the CCAs recipe for the induction of the slag-metal separation would be highly beneficial for the future industrial applicability of the process.

## References

[1] Colla V, Branca T A, Pietruck R, Wölfelschneider S, Morillon A, Algermissen D, Rosendahl S, Granbom H, Martini U and Snaet D 2023 Future Research and Developments on Reuse and Recycling of Steelmaking By-Products *Metals (Basel)* **13** 676

[2] Guézennec A-G, Huber J-C, Patisson F, Sessiecq P, Birat J-P and Ablitzer D 2005 Dust formation in Electric Arc Furnace: Birth of the particles *Powder Technol* **157** 2–11

[3] Barna R, Bae H-R, Méhu J, Sloot H van der, Moszkowicz P and Desnoyers C 2000 Assessment of chemical sensitivity of Waelz slag *Waste Management* **20** 115–24

[4] Branca T A, Colla V, Algermissen D, Granbom H, Martini U, Morillon A, Pietruck R and Rosendahl S 2020 Reuse and Recycling of By-Products in the Steel Sector: Recent Achievements Paving the Way to Circular Economy and Industrial Symbiosis in Europe *Metals (Basel)* **10** 345

[5] Mombelli D, Mapelli C, Barella S, Gruttadauria A and Di Landro U 2015 Laboratory investigation of Waelz slag stabilization *Process Safety and Environmental Protection* **94** 227–38

[6] Quijorna N, Coz A, Andres A and Cheeseman C 2012 Recycling of Waelz slag and waste foundry sand in red clay bricks *Resour Conserv Recycl* **65** 1–10

[7] Vegas I, Ibañez J A, San José J T and Urzelai A 2008 Construction demolition wastes, Waelz slag and MSWI bottom ash: A comparative technical analysis as material for road construction *Waste Management* **28** 565–74

[8] Antrekowitsch J, Rösler G and Steinacker S 2015 State of the Art in Steel Mill Dust Recycling *Chemie Ingenieur Technik* **87** 1498–503

[9] Khanapur N V., Tripathi B and Chandra T 2025 Incorporating Waelz slag to strengthen the properties of fine recycled aggregate concrete *Journal of Building Engineering* **104** 112235

[10] Grudinsky P I, Zinoveev D V., Dyubanov V G and Kozlov P A 2019 State of the Art and Prospect for Recycling of Waelz Slag from Electric Arc Furnace Dust Processing *Inorganic Materials: Applied Research* **10** 1220–6

[11] Fröhling M, Schwaderer F, Bartusch H and Schultmann F 2013 A Material Flow-based Approach to Enhance Resource Efficiency in Production and Recycling Networks *J Ind Ecol* **17** 5–19
[12] Rösler G, Pichler C, Antrekowitsch J and Wegscheider S 2014 "2sDR": Process Development of a Sustainable Way to Recycle Steel Mill Dusts in the 21st Century *JOM* **66** 1721–9
[13] Pichler C and Antrekowitsch J 2015 Recycling of Zinc- and Lead-Bearing Residues with Pyrolysis Gas *JOM* **67** 2038–46
[14] Pichler C and Antrekowitsch J 2017 Pyrolysis Gas as a Renewable Reducing Agent for the Recycling of Zinc- and Lead-Bearing Residues: A Status Report *JOM* **69** 999–1006
[15] Grudinsky P, Zinoveev D, Kondratiev A, Delitsyn L, Kulumbegov R, Lysenkov A, Kozlov P and Dyubanov V 2023 Reduction Smelting of the Waelz Slag from Electric Arc Furnace Dust Processing: An Experimental Study *Crystals (Basel)* **13** 318
[16] Zinoveev D V., Diubanov V G, Shutova A V. and Ziniaeva M V. 2015 Recycling of red muds with the extraction of metals and special additions to cement *Russian Metallurgy (Metally)* **2015** 19–21
[17] Mombelli D, Mapelli C, Barella S, Gruttadauria A, Ragona M, Pisu M and Viola A 2020 Characterization of cast iron and slag produced by red muds reduction via Arc Transferred Plasma (ATP) reactor under different smelting conditions *J Environ Chem Eng* **8** 104293
[18] Vitikka O, Iljana M, Heikkilä A, Tkalenko I, Kovtun O, Koriuchev N, Shehovsov D and Fabritius T 2024 Effect of Biocarbon Addition on Metallurgical Properties of Mill Scale-Based Auger Pressing Briquettes *ISIJ International* **64** ISIJINT-2023-417
[19] Dall'Osto G, Mombelli D, Scolari S and Mapelli C 2024 Role of the Biogenic Carbon Physicochemical Properties in the Manufacturing and Industrial Transferability of Mill Scale-Based Self-Reducing Briquettes *Metals (Basel)* **14** 882
[20] Dall'Osto G, Mombelli D, Trombetta V and Mapelli C 2024 Effect of Particle Size and Starch Gelatinization on the Mechanical and Metallurgical Performance of Jarosite Plus Blast Furnace Sludge Self-Reducing Briquettes *Journal of Sustainable Metallurgy* **10** 759–74
[21] Wang Y, Yang H, Zhang G, Kang J and Wang C 2020 Comprehensive recovery and recycle of jarosite residues from zinc hydrometallurgy *Chemical Engineering Journal Advances* **3** 100023
[22] Dall'Osto G, Scolari S, Mombelli D and Mapelli C 2025 Survey on the valorization of wood and agribusiness wastes for their application as fossil carbon substitutes in metallurgical processes *Biomass Convers Biorefin*
[23] Ullah A, Zhaoyun S, Tariq U, Uddin M I, Khatoon A and Rizvi S S 2022 Gray-Level Image Transformation of Paved Road Cracks with Metaphorical and Computational Analysis *Math Probl Eng* **2022** 1–14
[24] Ye Q, Li G, Peng Z, Lee J, Lin X, Rao M, Zhang Y and Jiang T 2019 Microwave-assisted self-reduction of composite briquettes of zinc ferrite and carbonaceous materials *Powder Technol* **342** 224–32
[25] Li Y, Ren J, Wei H, Khan I U, Omran M, Fabritius T, Shu Q, Wang C, Zang Y and Yu Y 2023 Reduction Kinetics of Cold-Bonded Briquette Prepared from Return Fines of Sinter with Carbon Monoxide and Coke *Steel Res Int* **94**
[26] Zulkania A, Rochmadi R, Hidayat M and Cahyono R B 2021 Reduction Reactivity of Low Grade Iron Ore-Biomass Pellets for a Sustainable Ironmaking Process *Energies (Basel)* **15** 137
[27] Castells B, Álvaro A G, Amez I, León D, Gascó A, Ortega M F and Hermosilla D 2024 Comparison of pyrolysis and anaerobic digestion processes for wheat straw biomass *Energy Reports* **12** 4390–401
[28] Coleti J L, Manfredi G V P, Vinhal J T, Junca E, Espinosa D C R and Tenório J A S 2020 Kinetic investigation of self-reduction basic oxygen furnace dust briquettes using charcoals from different biomass *Journal of Materials Research and Technology* **9** 13282–93
[29] Chai Y, Wang S, Zhang J, Shi Y, Liu P, An S, Wang C and Ning X 2024 Study on strength and reduction characteristics of iron ore powder-green carbon composite briquettes *Fuel* **377** 132741
[30] Fuwa T and Ban-Ya S 1969 Swelling of Iron Ore Pellets During Reduction *Transactions of the Iron and Steel Institute of Japan* **9** 137–47
[31] Michishita H and Tanaka H 2010 Prospects for coal-based direct reduction process *Kobelco Technology Review* **29** 69
[32] Higashi R, Maruoka D, Iwami Y and Murakami T 2024 Reduction and Carburization Behaviors of Iron Oxide Composite with Iron Carbide and Free Carbon *ISIJ International* **64** ISIJINT-2024-271
[33] Mishra S and Roy G G 2016 Effect of Amount of Carbon on the Reduction Efficiency of Iron Ore-Coal Composite Pellets in Multi-layer Bed Rotary Hearth Furnace (RHF) *Metallurgical and Materials Transactions B* **47** 2347–56
[34] Mombelli D, Mapelli C, Barella S, Gruttadauria A and Spada E 2019 Jarosite wastes reduction through blast furnace sludges for cast iron production *J Environ Chem Eng* **7** 102966
[35] Wei R, Cang D, Bai Y, Huang D and Liu X 2016 Reduction characteristics and kinetics of iron oxide by carbon in biomass *Ironmaking & Steelmaking* **43** 144–52

[36] Li H, Zhang H, Li K, Zhang J, Sun M and Su B 2020 Catalytic graphitization of coke carbon by iron: Understanding the evolution of carbon Structure, morphology and lattice fringes *Fuel* **279** 118531

[37] Chen Y-L, Chang J-E and Ko M-S 2017 Reusing Desulfurization Slag in Cement Clinker Production and the Influence on the Formation of Clinker Phases *Sustainability* **9** 1585

[38] Pliego-Cuervo Y B and Glasser F P 1977 The role of sulphates in cement clinkering reactions : Phase formation and melting in the system CaO-Ca2SiO4-CaSO4-K2SO4 *Cem Concr Res* **7** 477–81

[39] Queiroz C A R, Carvalho R J and Moura F J 2005 Oxidation of zinc sulphide concentrate in a fluidised bed reactor - Part 2: The influence of experimental variables on the kinetics *Brazilian Journal of Chemical Engineering* **22** 127–33

[40] Sabah S, Shahabuddin M, Rahbari A, Brooks G, Pye J and Rhamdhani M A 2024 Effect of gangue on CO2 emission for different decarbonisation pathways *Ironmaking & Steelmaking: Processes, Products and Applications* **51** 356–68

[41] Valenti F, Arcidiacono C, Chinnici G, Cascone G and Porto S M 2017 Quantification of olive pomace availability for biogas production by using a GIS-based model *Biofuels, Bioproducts and Biorefining* **11** 784–97

[42] World Steel Association, World Steel in Figures 2024

[43] Kaya M, Hussaini S and Kursunoglu S 2020 Critical review on secondary zinc resources and their recycling technologies *Hydrometallurgy* **195** 105362

[44] Mombelli D, Mapelli C, Dall'Osto G, Tavcar G, Kocjancic R, Radulovic D, Ristovic I, Ciszewski M, Kastivnik J and Mladenovic A 2024 EIT Raw Materials RIS-DUSTREC: investigation and selection of furnace dust samples for valuable metals recovery *La Metallurgia Italiana* **115** 8–23

[45] Grudinsky P, Yurtaeva A, Pankratov D, Pasechnik L, Musaelyan R and Dyubanov V 2024 The Waelz Slag from Electric Arc Furnace Dust Processing: Characterization and Magnetic Separation Studies *Materials* **17** 2224

Journal of Physics: Conference Series **3050** (2025) 012004 doi:10.1088/1742-6596/3050/1/012004

# Quantitative Phase Analysis in Carbon Steel EAF Slag for the Determination of Phase-Controlled Leaching Mechanism

**Davide Mombelli[1], Sara Scolari[1], Gianluca Dall'Osto[1], Carlo Mapelli[1]**

[1] Dipartimento di Meccanica, Politecnico di Milano, Via La Masa 1, 20156 Milano, Italy

*E-mail: sara.scolari@polimi.it

**Abstract.** The production of Electric Arc Furnace (EAF) slag is 150-180 kg per ton of steel, making its recycling essential for the steelmaking process's sustainability. Despite favourable mechanical properties as its good abrasion resistance, environmental challenges, such as the leaching of harmful elements (e.g., Cr, V, Ba and Mo), limit its direct reuse in industries like cement production, road construction and water filtration. The mineralogical composition of the slag is the key to address its leaching behaviour, as the hydration of specific phases (e.g., larnite and brownmillerite) can release these elements. Current regulations demand minimal environmental impact and require precise quantification of slag phases to identify and mitigate the leaching mechanisms. The slag is generally and rapidly characterized by means of X-Ray diffraction (XRD) and Scanning Electron Microscopy (SEM), however the quantitative analysis obtained by the Rietveld method is not always reliable because of the presence of preferential orientations common to several phases and the high sensitivity to the imposed background. Therefore, it is necessary to implement a methodology to refine the X-Ray analysis. Selective dissolution of larnite and brownmillerite, dangerous for Ba and Cr leaching, can be used for this purpose. In this work, different quantitative techniques were compared to tune-up the Rietveld analysis. These methods facilitated the accurate determination of critical phases like larnite and brownmillerite. SEM and XRD analyses have intrinsic limitation in quantifying phase contributions. In particular the identification of brownmillerite with XRD is overestimated respect the chemical prediction due to its strong crystallographic orientation al low 2θ angle. Selective chemical dissolution is instead the most reliable technique to quantify the phases, since a previous salicylic acid methanol (SAM) dissolution dissolves all the calcium silicates and a following potash sucrose (KOSH) treatment quantify all the calcium aluminates. The difficulties encountered doing only KOSH were associated to the formation of a viscous gel due to the hydration of silicates. If the dissolution of the brownmillerite was not complete, a second KOSH dissolution is required.

## 1. Introduction

Global steel industry has seen an intensification in production over the last 20 years, leading to higher formation of by-products as slag, dust and sludge. Slag plays a critical role in maintaining furnace operation protecting the molten bath from the atmosphere, keeping the temperature of the bath, ensuring metal purity, catching the inclusions and, finally, limiting wear of refractories and electrodes [1]. In 2023, 543.79 Mt of steel were produced in Electric Arc Furnace (EAF) [2], generating approximately 150-180 kg/$t_{steel}$ of slag [3].

According to annex 1 of Italian Ministerial Decree 05/02/98 the slag are defined as wastes with the 80% made of $SiO_2$, $Al_2O_3$, MgO and FeO [4]. However, far from being mere waste, slag are considered as an artificial hard rock due to their chemical and mineralogical composition and have found applications in various fields thanks to their physical and mechanical properties, such as abrasion and fragmentation resistance. They are often used in cement factories, in the construction and glass industries, in steel works and in primary and secondary smelting foundries for the recovery of ferrous and non-ferrous materials, or they can be used for the formation of embankments, road foundation and railway ballast, for environmental recovery and as a cover layer for municipal solid waste (MSW) landfills [4], [5].

However, severe limitations for slag reuse are imposed by each EU country because they suffer of volume instability, dusting effect and presence of potentially hazardous elements which can be leached into the environment upon hydration. The most concerning elements are Cr, V and Mo, originated from the scrap used as EAF charge, and Ba mainly contained in proler or in contaminated charged lime [6], [7].

The leaching process is significantly influenced by the chemical composition of the slag, its microstructure, granulometry, temperature and pH [8]. Among these factors, pH is the most important parameter [9], as the release of most constituents, governed by mineral dissolution or adsorption processes, is pH-dependent [10]. For instance, Król et al. [11] investigated the impact of environmental pH variations on the release of toxic metals from metallurgical slag, highlighting that the highest concentrations of leached elements occur in acidic conditions. Other studies instead demonstrate that the alkaline pH strongly affects the phenomena [7], [12]. The leaching behaviour is therefore primarily controlled by the leachate pH which is ruled by the slag chemical and mineralogical composition [7]. A logarithmic correlation has been identified between the larnite fraction and the leachate pH [12], as larnite largely determines the calcium concentration in the leachate and influences the final water pH. A higher larnite fraction in the slag corresponds to an increased ion concentrations (Al, Ca, and Ba) in the leachate, as confirmed by its direct proportionality with leachate electrical conductivity [12]. Thus, the presence of larnite and other calcium silicates is one of the limits to the recycling of slag since they react with water, may leading to Ba and V leaching. Other dangerous phases are calcium aluminate as brownmillerite and non-stoichiometric spinel which contribute to Cr release [13]. Additionally, brownmillerite promotes V leaching [5], [12], [13], [14]. Wustite, due to its low hydration rate, is instead not considered relevant in leaching phenomena, as it remains an insoluble phase [12], [13], [15]. Although spinels are generally known as stable phases [14], [16], an increase in slag basicity can lead to the release of V and Cr due to the formation of unstable ones [12].

Therefore, understanding the mechanisms behind element leaching and identifying the responsible phases is essential for allowing the reuse of slag [14]. The standard approach for mineralogical characterization relies on X-ray diffraction (XRD) combined with the Rietveld method for phase quantification. This technique is both fast and user-friendly when appropriate tools for structural validation are available [17]. However, its accuracy can be affected by several factors, including background noise, peak overlap, preferred grain orientation and the need to control multiple parameters, all of which can reduce the reliability of the quantification process. The presence of amorphous phases introduces broad humps in the diffraction patterns, making it difficult to determine the true background at high angles due to the high density of reflections. As a result, multiple re-evaluations are often required during refinement [18]. Peak overlap can affect the measurement, either due to broad peaks hiding others [17] or the presence of isostructural phases with identical crystalline patterns [19]. Another challenge arises when the background function fails to model the data accurately [18].

The structure determination process from XRD data begins with indexing, which involves defining the unit cell parameters. This is followed by space group determination and refinement of non-structural parameters. The phasing step is particularly critical, as it produces the initial crystal structure model. Subsequently, atom-related parameters are refined, with iterative adjustments between these final steps until the definitive model is obtained. Given the complexity of this process, it is time-consuming and requires significant user input, increasing the risk of human error [17].

Quantitative metallography through image analysis can be an alternative approach often followed in the field of metallurgy that includes, among other, the determination of specific phase fraction per unit volume [20], since it provides a rapid method for phase quantification by evaluating grayscale variations, pixel intensity correlations, and overall image homogeneity [21]. However, its effectiveness depends on sample size and heterogeneity. Increasing the sample size improves characterization, thus, a high number of particles are necessarily required to enhance accuracy. Moreover, the heterogeneity of the sample requires a statistically representative dataset through the evaluation of multiple images. Indeed, the sample analysed should be "representative" of the material [22]. Unfortunately, quantitative metallography based on image analysis provide the quantification of the phases in volume percentage, which then should be converted it into weight percentage. However, the density of the slag phases is not always tabulated in literature and consequently should be evaluated through different models [23] limiting the effectiveness of the procedure, and probably there is more going on than a simple volume to mass conversion.

Higher precision in determining specific mineral compounds may be reached by applying selective chemical dissolution (SCD) technique, that is a commonly employed means to use a range of chemical reagents to selectively extract mineral components from soils or sediments for the purposes of quantifying, separating, concentrating, or pre-cleaning of materials for analysis [24]. Thanks to a chemistry very close to that of cement and clinker [25], EAF slag could be selectively dissolved to defines the mass fraction of calcium silicates and calcium-iron aluminates, allowing direct comparison with the Rietveld refinement results. Calcium silicates contained in the slag may dissolve in acidic environmental following methodologies similar to those used for silicates extraction from clinker, while calcium aluminates require an alkaline environmental, a technique commonly applied in cement analysis [26], [27]. Unfortunately, the application of selective dissolution to EAF slag misses to the authors knowledge, making it necessary to establish an optimal sequence of treatments to maximize its potential in overcoming XRD limitations.

Thus, to address the limitations of the main quantitative analysis techniques and improve the accuracy of phase composition and abundance determination, this study proposes an integrated quantitative analysis protocol for EAF slag characterization. The approach compares and combines XRD with Rietveld refinement, image analysis and selective dissolution techniques to compensate for the criticalities of XRD alone and improve the accuracy of phase quantification.

## 2. Materials and Methods

The material studied in this work is a slag from an Italian electric arc furnace steelworks specialized in high quality carbon steel production, with a particle size, obtained through planetary ball milling, lower than 63 μm. The chemical composition was obtained through X-Ray fluorescence (XRF) and it is reported in Table 1.

**Table 1** Chemical composition of slag expressed in wt.%.

| $Fe_2O_3$ | CaO | $SiO_2$ | $Al_2O_3$ | MnO | MgO | $Cr_2O_3$ |
|---|---|---|---|---|---|---|
| 34.1 | 30.7 | 12.6 | 7.7 | 6.0 | 4.2 | 2.8 |

The mineralogical composition was determined using X-ray diffraction (XRD) by means of Rigaku SmartLab SE diffractometer in θ-θ configuration (Cu-Kα radiation: λ = 1.54 Å at 40 kV. 40 mA) to evaluate the mineralogy of the samples. The analysis was accomplished with a scan range of 5-90° 2θ through a 1D D/teX Ultra 250 detector featured by XRF suppression filter; the powdered materials were scanned at 1°/min with a step size of 0.02° and rotated at 60 rpm. Due to the specific lack of standards for mineralogical analyses by XRD, mineralogical analysis is based on the Crystallographic Open Database (COD), while the quantification of crystalline phases was carried out through Rietveld refinement.

The image analysis required the acquisition of 10 images through SEM-EDS. The quantification of each phase was then obtained analysing the grey-level of each map through the software ImageJ. The fraction of area in pixel ($A_{pixel}$) of the *i-sm* phase from each image was converted in mass fraction (*wt.%*) according to equation (1).

$$wt.\%_i = A_{i-pixel}\frac{\rho_i}{\rho_{slag}} \quad (1)$$

where $\rho_i$ is the density of the *i-sm* phase (larnite 3.3 g/cm$^3$, brownmillerite 3.8 g/cm$^3$, spinel 4.8 g/cm$^3$ [28] and wustite 4.7 g/cm$^3$ according to Mills model [23]) and $\rho_{slag}$ is the density of the slag (3.7 g/cm$^3$ according to Mills model [23]).

According to the experimental procedures already defined for cements [26], [29], [30], calcium silicates were dissolved using a salicylic acid-methanol (SAM) solution, while calcium aluminates and iron phases were removed using a potassium hydroxide-sucrose (KOSH) solution. After mixing slag and solution for a determined time span, 5 min more were waited to allow a primary mixture decantation before its vacuum-filtering through a PTFE filter membrane. The residual solution was continued stirred to improve the filtration avoiding the complete decantation of the solid particles. The solid residues collected on the filter was dried for 2 h at 65 °C and weighted. It was then re-analysed via XRD to assess phase depletion and verify the initial quantification. The experimental conditions for the preparation of the two solutions are reported in Table 2 and the dissolutions performed on the slag are listed as follows:

- SAM dissolution
- KOSH dissolution
- Combined KOSH+SAM dissolution (the second dissolution was performed on the residues of the first)
- Combined SAM+KOSH dissolution (the second dissolution was performed on the residues of the first)

**Table 2** Experimental condition for solutions preparation.

| Dissolution type | Materials | Temperature | Filter type |
|---|---|---|---|
| SAM | 300 mL methanol<br>20 g salicylic acid<br>5 g slag | 25 °C | Hydrophobic (mesh size 0.45 μm) |
| KOSH | 15 g potassium hydroxide<br>15 g sucrose<br>150 mL distilled water<br>5 g slag | The materials were heated at 90 °C before adding the slag | Hydrophilic (mesh size 0.45 μm) |

## 3. Results and discussion

The XRD pattern of the untreated slag, reported in Figure 1, identified four crystalline compounds commonly found in an EAF slag [12], [13], [14], [16], namely wustite (FeO), larnite ($2CaO{\cdot}SiO_2$), brownmillerite ($4CaO{\cdot}Al_2O_3 \cdot Fe_2O_3$) and a Cr-rich spinel ($(Mg,Mn,Fe)(Al,Mn,Fe,Cr)_2O_3$). Larnite was the only phase identified by the XRD which contains $SiO_2$, therefore in this specific case the maximum theoretical amount of larnite was calculated starting from the average chemical composition and resulted equal to 36.2 wt.%. Same quantification (36.2 wt.% of larnite) was obtained performing the Rietveld refinement, which also identified 22.1% wustite, 22.6 wt.% brownmillerite and 19.1 wt.% spinel. Under the hypothesis that all the residual CaO not yet bound in larnite goes into brownmillerite, its fraction was computed from average chemical composition and resulted equal to 15.7 wt.%. This time, Rietveld refinement overestimated brownmillerite amount of approximately 44%, probably due to the preferred orientation of the phase that gave high intensity to the peak at 12 °2θ (Figure 1).

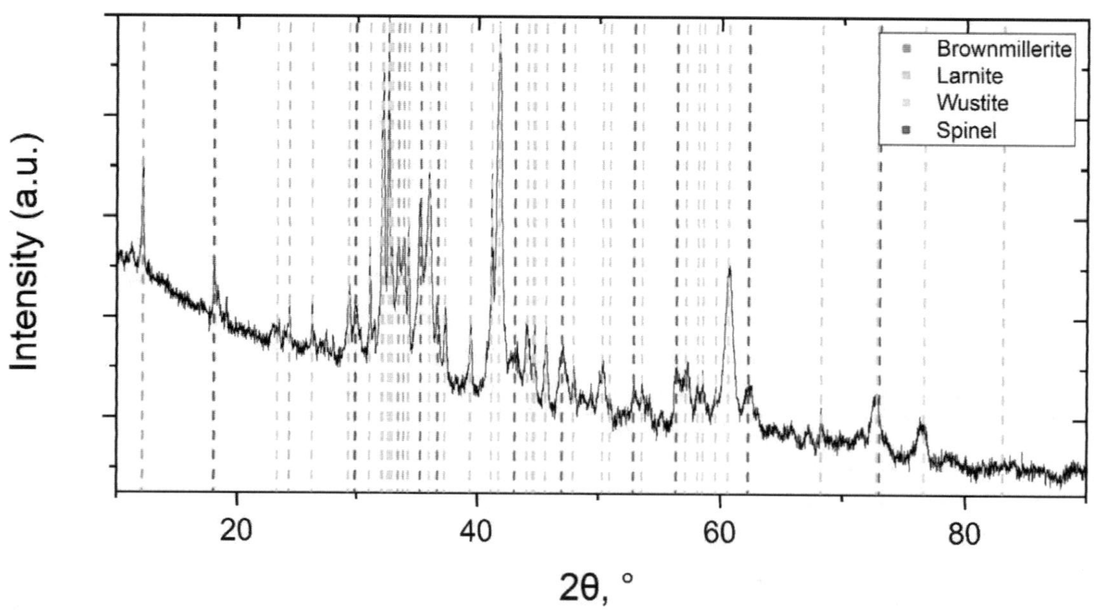


**Figure 1** XRD pattern of untreated slag.

### *3.1 Image analysis*

The SEM-EDS map (Figure 2) shows the presence of the same phases identified in the XRD pattern. The brownish zones identified by letter *"L"* were larnite being rich in calcium and silicon, the magenta zones "B" identified brownmillerite being rich in aluminium, iron and calcium. The light blue areas *"W"* are wustite having high iron content, while the green ones *"C"* are Cr-rich spinel. Silicon distribution confirmed the hypothesis that $SiO_2$ is only bound with CaO in larnite, while CaO distributes effectively only in larnite and brownmillerite, allowing for the determination of the latter from chemical composition.

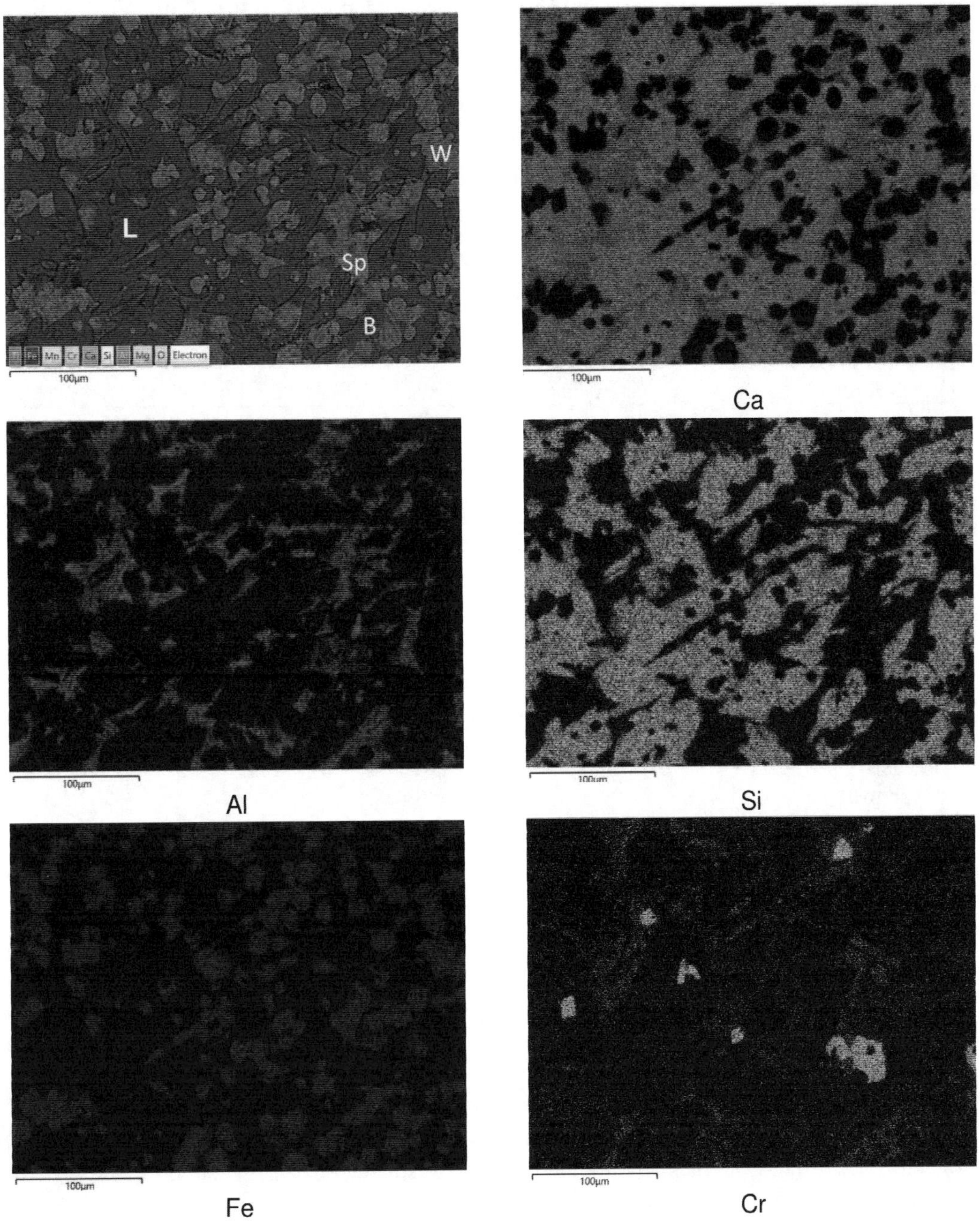


**Figure 2** SEM-EDS map of untreated slag. (W: wustite; B: brownmillerite; C: chromite; L: larnite).

The results of the image analysis conducted by ImageJ software are reported in Table 3. The amount of brownmillerite was similar to the Rietveld refinement, but further overestimated the chemical calculation. On the contrary, the 11% reduction in larnite observed in the SEM-EDS evaluation may be attributed to its dissolution during sample preparation or heterogeneity of the material. The increase in wustite and chromite percentages (15 and 17%, respectively) could be instead associated to difficulties in their identification during the image analysis. The heterogeneity of the slag and the limited number of pictures evaluated were not only the main limitation in the quantification of such phases, but they are also difficult to identify since they have similar grey-level in back-scattered images. From these results, it is quite clear that both Rietveld refinement and image analysis suffers from some bias and errors that cannot ensure a reliable estimation of the real phases' amount, especially if the chemical composition is used as reference. Thus, SCD seems a more than viable way to get precisely information about actual quantities, at least for such compounds that can be selectively leached out.

**Table 3** Quantification of each phase expressed in wt.% of image analysis and Rietveld refinement.

| Phase | Image analysis [wt.%] | Rietveld [wt.%] | Chemical computation [ wt.%] |
|---|---|---|---|
| Wustite | 25.4 ± 3.8 | 22.1 | - |
| Larnite | 32.9 ± 2.6 | 36.2 | 36.2 |
| Brownmillerite | 23.3 ± 3.8 | 22.6 | 15.7 |
| Spinel | 15.8 ± 6.3 | 19.1 | - |

*3.2 SAM dissolution*

Comparing the quantification of the phases through the different methods (Table 4), all the techniques appeared reliable to quantify the calcium silicate, since the results had the same order of magnitude.

**Table 4** Quantification of each phase expressed in wt.% of slag after SAM dissolution and pre-treated slag.

| Phase | Dissolution SAM [wt.%] | Image analysis [wt.%] | Rietveld pre-treated [wt.%] | Chemical computation [ wt.%] |
|---|---|---|---|---|
| Wustite | | 25.4 ± 3.8 | 22.1 | |
| Larnite | 33.2 | 32.9 ± 2.6 | 36.2 | 36.2 |
| Brownmillerite | | 23.3 ± 3.8 | 22.6 | 15.7 |
| Spinel | | 15.8 ± 6.3 | 19.1 | |

However, the mass fraction dissolved using SAM solution was 8.3% lower than the maximum theoretical fraction of larnite identified by the chemical computation. The XRD was therefore performed on the post-treated slag to verify the dissolution of the calcium silicate. Figure 3 shows the total disappearance of the characteristic peaks associated to larnite, confirming the complete dissolution of the phase and the suitability to use the SAM dissolution as a technique to quantify it. The discrepancy reported in Table 4 was therefore associated to the heterogeneity of the slag. This was then demonstrated by repeating the procedure on three more samples prepared by the same slag batch and the results ranged from 31.7 to 34.7 wt.%, with an average value on 33.2 wt.%.

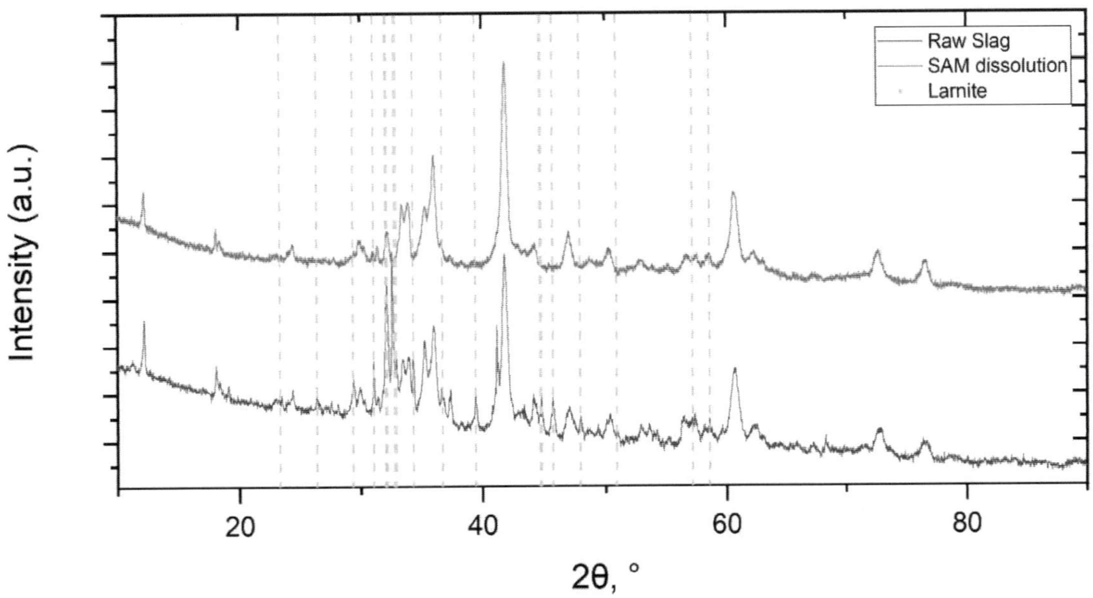


**Figure 3** XRD pattern comparison of untreated slag and post SAM dissolution.

*3.3 KOSH dissolution*

Considering the results got with the different techniques and reported in Table 5, the KOSH dissolution was not a suitable solution to quantify brownmillerite because a residual amount remained undissolved in the slag. The mass fraction dissolved during the KOSH dissolution of the raw slag indeed was about 12.3%, which is 46% lower than the amount of brownmillerite assumed with the Rietveld refinement and 28% lower than the maximum theoretical brownmillerite available in the slag from chemistry.

**Table 5** Quantification of each phase expressed in wt.% of slag after KOSH dissolution and pre-treated slag.

| Phase | Dissolution KOSH [wt.%] | Image analysis [wt.%] | Rietveld pre-treated [wt.%] | Chemical computation [ wt.%] |
|---|---|---|---|---|
| Wustite | | 25.4 ± 3.8 | 22.1 | |
| Larnite | | 32.9 ± 2.6 | 36.2 | 36.2 |
| Brownmillerite | 12.3 | 23.3 ± 3.8 | 22.6 | 15.7 |
| Spinel | | 15.8 ± 6.3 | 19.1 | |

The Rietveld overestimated the phase respect the chemical computation, but also the latter predicted a maximum theoretical amount of brownmillerite greater than the fraction dissolved. Thus, the partial dissolution of the phase was investigated through XRD. As expected, Figure 4 shows the peaks of the phase still present in the XRD patter of the slag after the treatment, confirming the non-complete reliability on KOSH dissolution to quantify the calcium aluminate.

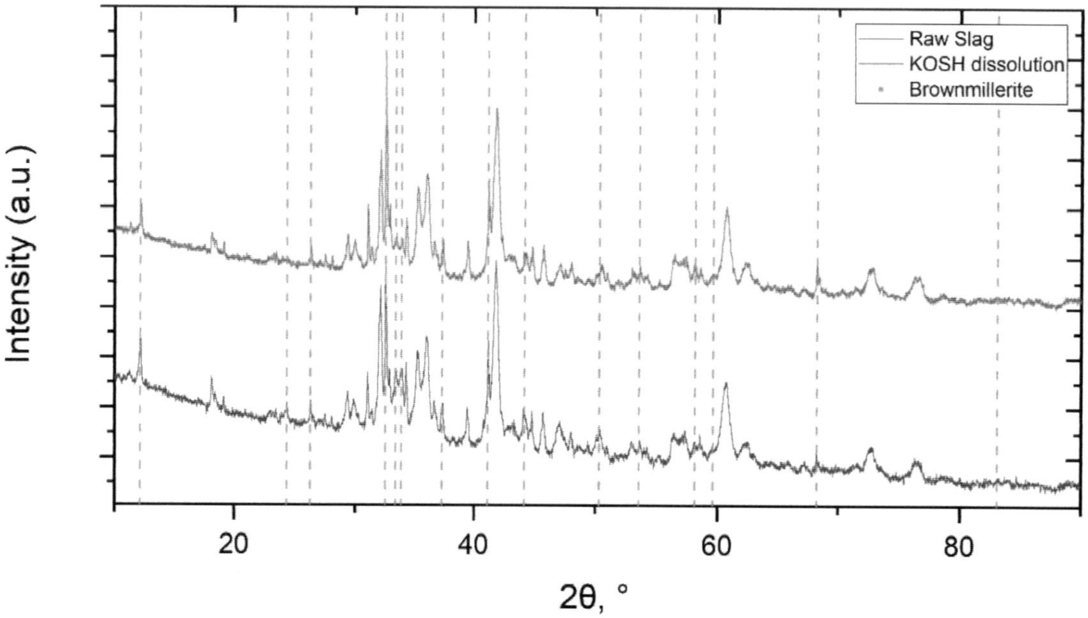


**Figure 4** XRD pattern comparison of untreated slag and post KOSH dissolution.

The partial dissolution of the phases containing CaO, $Al_2O_3$, $Fe_2O_3$ and MgO in KOSH was already reported by Li et al [31], who explained that the 4.22 wt.% of MgO (Table 1) in the slag could inhibit the dissolution of calcium and calcium iron aluminates. In addition, the hydration of the silicates due to environmental alkalinity and water presence led to the formation of a viscous gel [32], [33] which obstructed the filter limiting the filtration. The sucrose in the solution further increased the viscosity of the gel due to the favoured hydration [34]. Thus, the dissolution of the calcium silicates, responsible for hydrated gel phase, is reasonable to be conducted before applying for KOSH treatment, if the goal is to enhance brownmillerite and other aluminates leaching. This is the reason why the combined KOSH-SAM dissolution was not performed on the sample as originally scheduled.

*3.4 SAM-KOSH dissolution*

The SAM-KOSH dissolution consisted in same sample subject to the SAM dissolution followed by KOSH. The treatment in SAM led to an average mass dissolution of 33.2 wt.% as already discussed, while the following KOSH dissolution led to an absolute loss of 21.1 wt.% of the SAM-treated slag mass. The reliability of the procedure was confirmed through other three repetitions for which the mass loss ranged from 19.4 to 22.6 wt.%.

The quantification of each phase through the different techniques is reported in Table 6. The fraction dissolved by the KOSH dissolution calculated on the raw sample was 14.1 wt.%, 10% lower than the maximum theoretical amount of brownmillerite in the slag, according to the chemical computation.

**Table 6** Quantification of each phase expressed in wt.% of slag after SAM-KOSH dissolution and pre-treated slag.

| Phase | Dissolution SAM [wt.%] | Dissolution SAM-KOSH [wt.%] | Image analysis [wt.%] | Rietveld pre-treated [wt.%] | Chemical computation [ wt.%] |
|---|---|---|---|---|---|
| Wustite | | | 25.4 ± 3.8 | 22.1 | |
| Larnite | 33.2 | | 32.9 ± 2.6 | 36.2 | 36.2 |
| Brownmillerite | | 14.1 | 23.3 ± 3.8 | 22.6 | 15.7 |
| Spinel | | | 15.8 ± 6.3 | 19.1 | |

To verify the dissolution efficiency of the calcium aluminates, XRD pattern of the pre- and post-treated slag are compared in Figure 5. Although the main peak of brownmillerite at 12° 2θ was not more present, peaks around 33 °2θ suggested some residual amount and confirmed the not complete dissolution of brownmillerite.

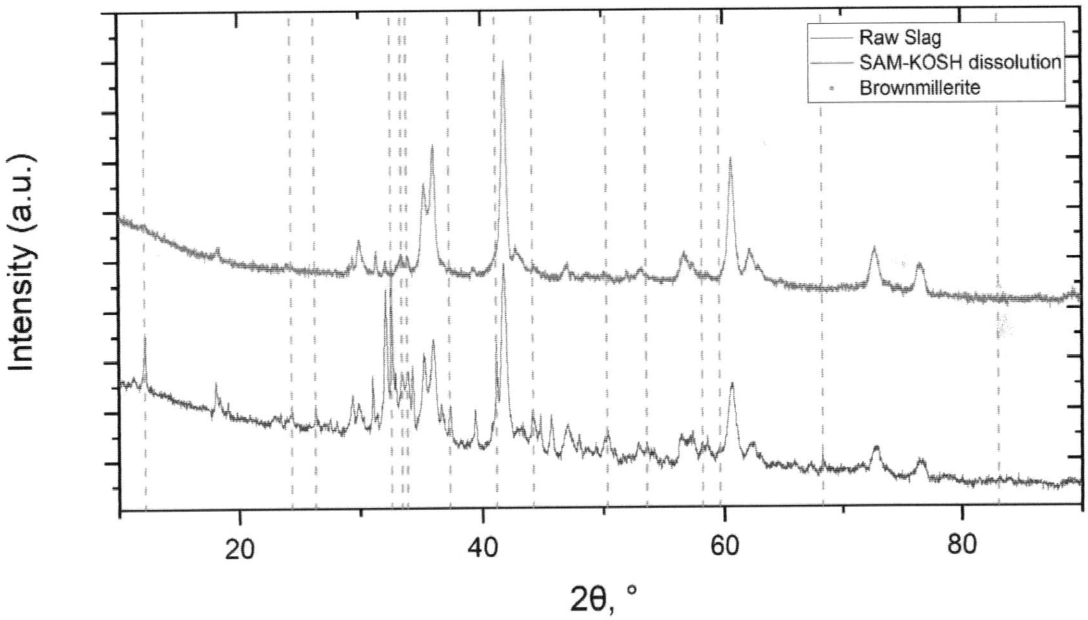


**Figure 5** XRD pattern comparison of untreated slag and post SAM-KOSH dissolution.

To complete dissolve the calcium aluminate, a further KOSH dissolution was performed on the sample already treated, leading to an absolute mass loss of 7.0 wt.% (Table 7). Thus, the total amount of brownmillerite dissolved during the double KOSH dissolution followed to the SAM was 17.3 wt.%, even exceeding the chemical computation. Although during samples preparation, quartering and homogenisation were carried out, the reason behind this discrepancy can relies on the heterogeneity of the slag.

**Table 7** Quantification of each phase expressed in wt.% of slag after SAM-KOSH-KOSH dissolution and pre-treated slag.

| Phase | Dissolution SAM [wt.%] | Dissolution SAM-KOSH [wt.%] | Dissolution SAM-KOSH-KOSH [wt.%] | Image analysis [wt.%] | Rietveld pre-treated [wt.%] | Chemical computation [ wt.%] |
|---|---|---|---|---|---|---|
| Wustite | | | | 25.4 ± 3.8 | 22.1 | |
| Larnite | 33.2 | | | 32.9 ± 2.6 | 36.2 | 36.2 |
| Brownmillerite | | 14.1 | 3.2 | 23.3 ± 3.8 | 22.6 | 15.7 |
| Chromite | | | | 15.8 ± 6.3 | 19.1 | |

To verify the complete dissolution, XRD was performed on the sample and as shown in Figure 6, the characteristic peaks of brownmillerite disappeared, confirming the capability of the double treatment to dissolve all the phase.

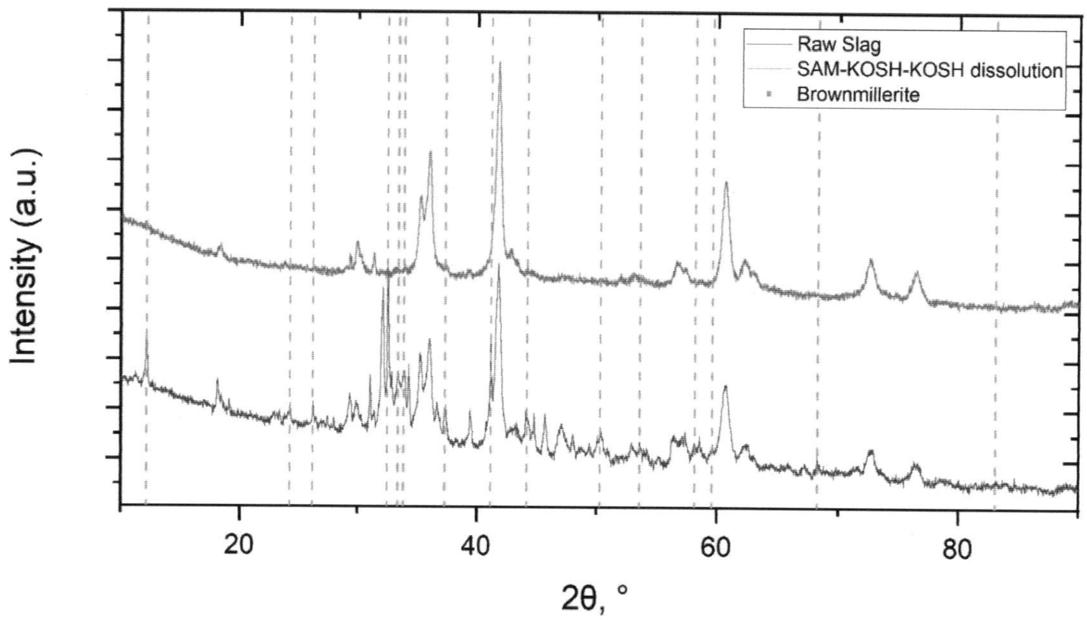


**Figure 6** XRD pattern comparison of untreated slag and post SAM-KOSH-KOSH dissolution.

The combined SAM-KOSH-KOSH dissolution was therefore the most reliable technique to quantify both larnite and brownmillerite for the slag analysed. The previous removal of the larnite allowed a following better dissolution of the brownmillerite because it was avoided the hydration of the silicate and the consequent formation of the gel which obstructed the filtration. The best protocols to define the phase quantification hence leads to a first dissolution in SAM solution, to remove all the calcium silicates, and a following KOSH dissolution to leach all the calcium aluminates. If the dissolution of the latter was still partial, a second KOSH dissolution is suggested.

*3.5 Validation of the protocols with other samples*

The protocol defined performing the selective dissolution of the slag suggested the combined SAM-KOSH technique. The validation of this procedure was done analysing 5 other samples

Journal of Physics: Conference Series **3050** (2025) 012004 doi:10.1088/1742-6596/3050/1/012004

(origin and chemical composition listed in Table 8) and compared with the base case. All the samples' characterization was already performed and available in cited references.

**Table 8** Chemical composition of other slag expressed in wt.%.

| Sample | Origin | MgO | $Al_2O_3$ | $SiO_2$ | CaO | $Fe_2O_3$ | MnO | $Cr_2O_3$ | Reference |
|---|---|---|---|---|---|---|---|---|---|
| A1 | Rebar steel | 3.2 | 9.0 | 12.4 | 26.1 | 40.8 | 6.2 | 4.1 | [13], [35] |
| A2 | Rebar steel | 4.3 | 7.4 | 10.4 | 27.7 | 41.4 | 6.7 | 2.7 | [13], [35] |
| C | High-alloyed steel | 8.0 | 3.5 | 9.5 | 29.8 | 32.2 | 5.3 | 11.7 | [36] |
| F | High-alloyed steel | 3.0 | 4.4 | 10.0 | 19.4 | 42.8 | 7.6 | 14.0 | [12] |
| S | Quality steel | 5.5 | 7.4 | 8.1 | 26.4 | 41.7 | 6.9 | 2.4 | [37] |

The numerical computation of the larnite was not possible in case of sample A1 because silicon was also contained in gehlenite ($Ca_2Al(SiAl)O_7$). The same problem happened with sample C because the $SiO_2$ was also present in a minor phase, scheelite. The quantification of the phases based on the chemical composition is indeed not always feasible since the same element could form different compounds limiting the possibility to apply a numerical calculation. Hence it could not be considered as a suitable procedure to quantify the phases inside all the slag, although it can give an indication about the reliability of other methods. It was instead possible to quantify larnite in sample F and also brownmillerite in samples A2 and S. The results are reported in Table 9.

**Table 9** Chemical computation of phases inside slag expressed in wt.%.

| Sample | Larnite [wt.%] | Brownmillerite [wt.%] |
|---|---|---|
| A2 | 29.8 | 18.0 |
| F | 28.7 | |
| S | 23.1 | 24.5 |

The SAM dissolution was performed in all the samples because they all contained larnite according to the XRD analysis, while the KOSH dissolution was performed only on A2, C and S since they are the only slag which contained brownmillerite.

The phase quantification on the untreated slag performed through Rietveld refinement and the dissolved fraction got from the selective dissolution are reported in Table 10.

**Table 10** Rietveld on pre-treated slag and fraction dissolved in SAM and SAM-KOSH dissolution expressed in wt.%. [n.f. means not functioning].

| Sample | Wustite | Chromite | Larnite | Brownmillerite | Spinel | Gehlenite | SAM dissolution | SAM-KOSH dissolution | SAM-KOSH-KOSH dissolution |
|---|---|---|---|---|---|---|---|---|---|
| A1 | 27.7 | - | 38.8 | - | 13.2 | 20.3 | 36.6 | - | - |
| A2 | 36.0 | - | 32.8 | 23.2 | 8.0 | - | 30.7 | 16.7 | - |
| C | - | 46.5 | 21.3 | 27.9 | - | - | 21.2 | n.f. | - |
| F | 33.2 | 29.6 | 25.4 | - | 11.7 | - | 25.0 | - | - |
| S | 26.7 | - | 26.8 | 29.2 | 17.2 | - | 26.6 | 17.1 | 2.4 |

As already happened in the base case, the quantification of larnite though the three different methods gave very similar amount. The quantification of brownmillerite through selective dissolution resulted instead more critical since it was possible only on samples A2 and S. The KOSH dissolution of A2 led to an absolute mass loss of 24.0 wt.% of the already treated slag which is 16.7 wt.% of the raw material, value close to the amount calculated with the chemical computation. Once again, Rietveld refinement overestimated the amount of the calcium aluminate probably due to the high intensity of the peaks at 12 °2θ caused by the strong orientation of the phase. The sample S dissolved 23.3 wt.% of the SAM-treated slag and 4.2 wt.% on the second KOSH dissolution leading to a total mass loss of 19.5 wt.% of the raw slag. Interestingly, the behaviour of S sample was very similar to the base case, namely it required two stages of KOSH for a total removal of brownmillerite from the sample. In addition, the experimental determination of brownmillerite on S samples largely differ from the computed amount of approximately 25%. This discrepancy may be due to the fact that, by looking at the brownmillerite at high magnification, a Ti-rich phase was identified (Figure 7, Table 11). This means that not all the residual CaO that was not bound in larnite was available for brownmillerite, but part of this CaO was bound with $TiO_2$ to form perovskite. This would result in a lower amount of computed brownmillerite. Unfortunately, $TiO_2$ concentration was not provided by XRF. Hence, it is not possible to account for.

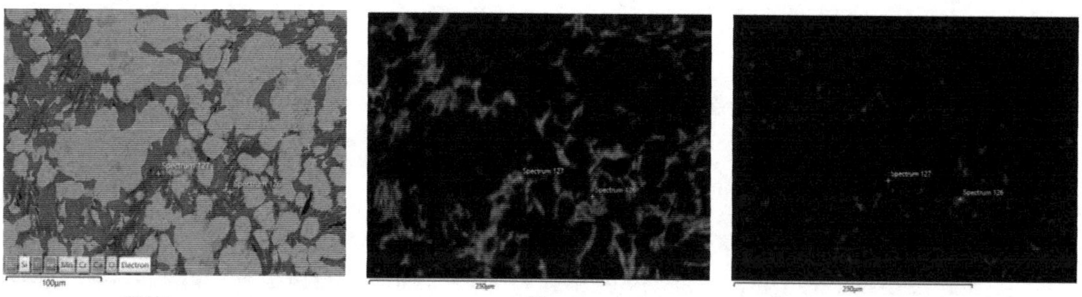

EDS map Al Ti

**Figure 7** EDS map, Al and Ti elemental maps for sample S with local EDS analysis identification point.

**Table 11** SEM-ESD spectrum expressed in wt.% for the point located in Figure 7.

| # | Al | Ca | Fe | Si | Ti | phase |
|---|---|---|---|---|---|---|
| 126 | 13.4 | 45.4 | 18.1 | 2.8 | 16.7 | Brownmillerite/Perovskite |
| 127 | 31.2 | 35.7 | 21.5 | 2.4 | - | Brownmillerite |

On the contrary, in sample C the high concentration of Cr bounded to brownmillerite modified its chemical composition and hindered its dissolution. Considering all the sample, the progressive increase of Cr in chemical composition and in particular its concentration identified through EDS point analysis in the brownmillerite led to limitation in the phase dissolution.

Samples with low Cr as A2 (0.29-0.74 wt.%) allowed the phases quantification after the combined SAM-KOSH dissolution, while sample S and the sample used to define the protocol (base case) required instead a double KOSH dissolution containing 3.44 and 4.26 wt.% of Cr, respectively, in the calcium aluminate. The dissolution was instead not possible in case of high-Cr slag as C (8.27-18.81 wt.% in brownmillerite).

The selective dissolution protocol defined for the base case was validated performing the selective dissolution and the Rietveld refinement on 5 other samples. Figure 8a shows that both the quantitative techniques provided the amount of larnite close the maximum theoretical one calculated according to the chemistry. The discrepancies were associated to the intrinsic heterogeneity of the slag. It is worthy to mention that from the results, Rietveld refinement tends to slightly overestimate the larnite fraction in comparison to the SCD technique. Figure 8b shows instead a strong overestimation of the brownmillerite with the Rietveld refinement, while the KOSH dissolution gave the same quantification of the chemical computation. This effect can be accounted for both the strong orientation of the peak at 12 °2θ and the difficulties to define the background at low 2θ angles. Further samples should be analysed to unequivocally demonstrate this behaviour.

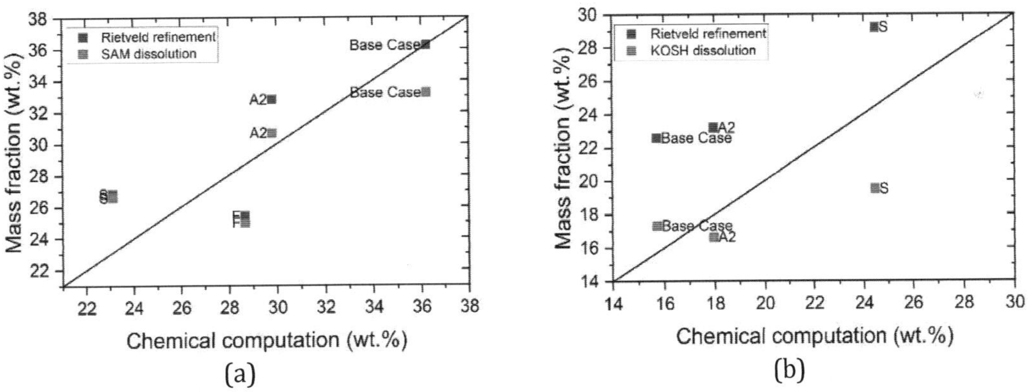


**Figure 8** Comparison between phases quantification with different methods for different slag.

### *3.6 Correlation between phases amount and leaching beahviour*

The accurate quantification of larnite and brownmillerite is important to allow the recycle of the slag. The two phases are indeed responsible for the leaching of harmful elements, especially Ba and V [5], [12], [13], [14]. To underline the dual relationship between V and Ba and the amount of the phases under exam, Figure 9 shows the ratio between the calcium silicate and the calcium aluminate and the concentration of harmful elements in the leachate.

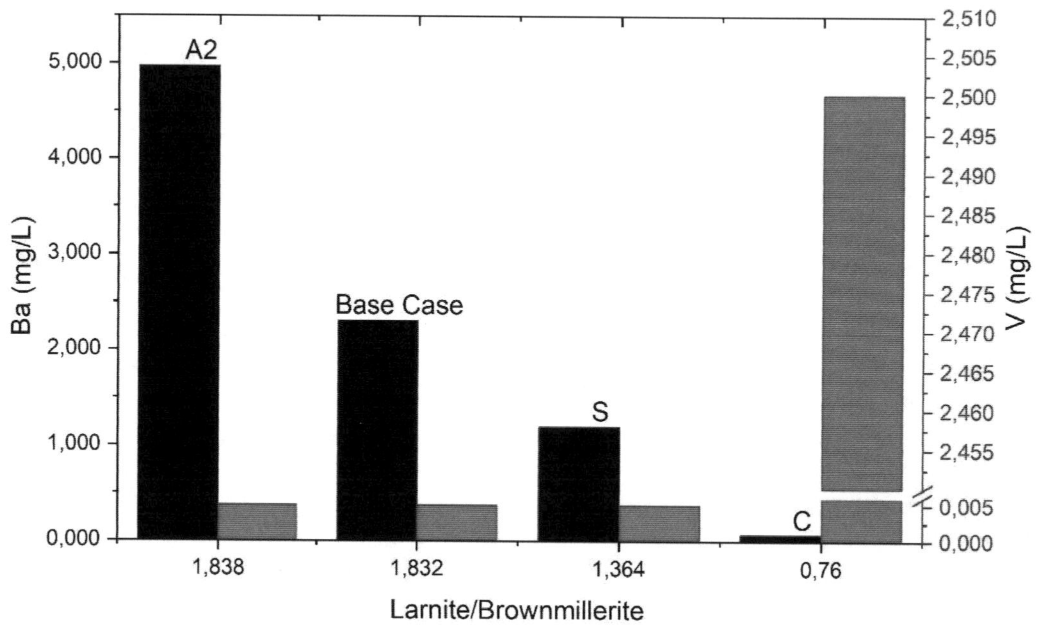


**Figure 9** Amount of Ba and V leached, expressed in mg/L, according to the ratio between larnite and brownmillerite dissolved in the slag.

The amount of the leached element is available in the following references, [12], [13], [35], [36], [37], [38] Base case and samples A2 and S released less than 0.005 mg/L of V, namely, below the sensitivity of the instrument.

The correlation highlighted that if there is Ba, the leaching of V does not take place and if there is leached V, the leaching of Ba is limited. In addition, increasing the ratio between larnite and brownmillerite, the amount of Ba increased, while decreasing the ratio, the amount of V increased in turn. This evidence confirmed the hypothesis done by Mombelli et al. which correlated the release of Ba with the calcium silicate and the leaching of V with the calcium aluminate [12], highlighting the importance to define a correct and accurate methodology for the phases' quantification in EAF slag.

## 4. Conclusions

In order to solve the limitation of phase quantification by Rietveld refinement in EAF slag, image analysis and selective dissolution were performed and the results compared with the numerical computation based on the chemistry. The results are summarised as follows:

- The Rietveld refinement is reliable in the quantification of larnite but the preferred orientation of the brownmillerite overestimated it respect the chemical computation.
- The image analysis has limitation in the quantification of the phases due to heterogeneity of the slag, similar grey-level of iron-rich phases and issues in defining the actual density of the compounds.
- The SAM dissolution is a reliable technique to quantify larnite, as it worked on six different slag samples both from chemistry and production process.
- The dissolution of brownmillerite does not work performing only KOSH dissolution, a previous SAM treatment is requested to avoid the silicates hydration.

- According to the amount of Cr in the brownmillerite the protocols may change. If the amount is low (< 1 wt.%), combined SAM-KOSH dissolution is sufficient, increasing the Cr amount a double KOSH treatment is required until a high amount of Cr (> 8 wt.%) inhibits the dissolution at all.

## Acknowledgment

The author would like to acknowledge the students for the collaboration in the experimental campaign, Marta Balconi, Thi Dung Giada Barzaghi, Matilde Danielli, Tommaso Marconi, Simone Mazzeo, Eleonora Moretti, Silvia Ida Elisa Montemurro, Marco Parisi, Dongyang Davide Fang, Leonardo Quagliata, Niccolò Cuocolo and Stefano Brandi.

## References

[1] W. Nicodemi and C. Mapelli, *Siderurgia*. Milano: Associazione Italiana di Metallurgia, 2011.

[2] "World Steel Association," 2023.

[3] A. Piemonti, A. Conforti, L. Cominoli, A. Luciano, G. Plizzari, and S. Sorlini, "Exploring the Potential for Steel Slags Valorisation in an Industrial Symbiosis Perspective at Meso-scale Level," *Waste Biomass Valorization*, 2022, doi: 10.1007/s12649-022-01940-5.

[4] "Decreto ministeriale del 5 febbraio 1998: Individuazione dei rifiuti non pericolosi sottoposti alle procedure semplificate di recupero ai sensi degli articoli 31 e 33 del decreto legislativo 5 febbraio 1997, n. 22.", Gazzetta Ufficiale n°88 del 16 aprile 1998- Supplemento Ordinario."

[5] S. Barella, A. Gruttadauria, F. Magni, C. Mapelli, and D. Mombelli, "Survey about safe and reliable use of EAF slag," *ISIJ International*, vol. 52, no. 12, pp. 2295–2302, 2012, doi: 10.2355/isijinternational.52.2295.

[6] D. Mombelli *et al.*, "Analysis of Electric Arc Furnace Slag (SRI)," *Steel research*, vol. 83, no. 11, pp. 1012–1019, 2012.

[7] D. Mombelli, C. Mapelli, S. Barella, C. Di Cecca, G. Le Saout, and E. Garcia-Diaz, "The effect of chemical composition on the leaching behaviour of electric arc furnace (EAF) carbon steel slag during a standard leaching test," *J Environ Chem Eng*, vol. 4, no. 1, pp. 1050–1060, Mar. 2016, doi: 10.1016/j.jece.2015.09.018.

[8] I. Strandkvist, B. Björkman, and F. Engström, "Synthesis and dissolution of slag minerals - a study of β-dicalcium silicate, pseudowollastonite and monticellite," *Canadian Metallurgical Quarterly*, vol. 54, no. 4, pp. 446–454, Oct. 2015, doi: 10.1179/1879139515Y.0000000022.

[9] K. Komonweeraket, B. Cetin, A. H. Aydilek, C. H. Benson, and T. B. Edil, "Effects of pH on the leaching mechanisms of elements from fly ash mixed soils," *Fuel*, vol. 140, pp. 788–802, Jan. 2015, doi: 10.1016/j.fuel.2014.09.068.

[10] "H. A. van der Sloot and J. J. Dijkstra, 'Development of horizontally standardized leaching tests for construction materials: a material based or release based approach?,' 2004."

[11] A. Król, K. Mizerna, and M. Bożym, "An assessment of pH-dependent release and mobility of heavy metals from metallurgical slag," *J Hazard Mater*, vol. 384, p. 121502, Feb. 2020, doi: 10.1016/j.jhazmat.2019.121502.

[12] D. Mombelli, C. Mapelli, S. Barella, C. Di Cecca, G. Le Saout, and E. Garcia-Diaz, "The effect of microstructure on the leaching behaviour of electric arc furnace (EAF) carbon steel slag," *Process Safety and Environmental Protection*, vol. 102, pp. 810–821, Jul. 2016, doi: 10.1016/j.psep.2016.05.027.

[13] D. Mombelli, C. Mapelli, S. Barella, A. Gruttadauria, G. Le Saout, and E. Garcia-Diaz, "The efficiency of quartz addition on electric arc furnace (EAF) carbon steel slag stability," *J Hazard Mater*, vol. 279, pp. 586–596, Aug. 2014, doi: 10.1016/j.jhazmat.2014.07.045.

[14] I. Strandkvist *et al.*, "Minimizing Chromium Leaching from Low-Alloy Electric Arc Furnace (EAF) Slag by Adjusting the Basicity and Cooling Rate to Control Brownmillerite Formation," *Applied Sciences*, vol. 10, no. 1, p. 35, Dec. 2019, doi: 10.3390/app10010035.

[15] D. Mombelli, S. Barella, A. Gruttadauria, C. Mapelli, G. Le Saout, and E. Garcia-Diaz, "Effects of Basicity and Mesh on Cr Leaching of EAF Carbon Steel Slag," *Applied Sciences*, vol. 9, no. 1, p. 121, Dec. 2018, doi: 10.3390/app9010121.

[16] S. Neuhold *et al.*, "Investigation of Possible Leaching Control Mechanisms for Chromium and Vanadium in Electric Arc Furnace (EAF) Slags Using Combined Experimental and Modeling Approaches," *Minerals*, vol. 9, no. 9, p. 525, Aug. 2019, doi: 10.3390/min9090525.

[17] Giada Truccolo, "Strengths and Limitations of Powder X-ray Diffraction," 2023.

[18] Christian Baerlocher and Lynne B. McCusker, "The potential difficulty is that the background may not be well-modelled by the function used.," in *Studies in Surface Science and Catalysis*, 1994.

[19] J. Vereš, V. Šepelálk, and S. Hredzák, "Chemical, Mineralogical and Morphological Characterisation of Basic Oxygen Furnace Dust," *Transactions of the Institutions of Mining and Metallurgy, Section C: Mineral Processing and Extractive Metallurgy*, vol. 124, no. 1, pp. 1–8, Mar. 2015, doi: 10.1179/1743285514Y.0000000069.

[20] George Vander Voort, *Introduction to Quantitative Metallography*, vol. 1. Buehler Ltd., 2005.

[21] O. Khomiak, J. Benndorf, and G. Verbeek, "Sub-Surface Soil Characterization Using Image Analysis: Material Recognition Using the Grey Level Co-Occurrence Matrix Applied to a Video-CPT-Cone," *Mining*, vol. 4, no. 1, pp. 91–105, Feb. 2024, doi: 10.3390/mining4010007.

[22] OCCHIO SA, "PARTICLE CHARACTERIZATION BY IMAGE ANALYSIS ."

[23] "K. Mills, 'The estimation of slag properties', Short course presented as part of Southern African Pyrometallurgy 2011, 7 March 2011."

[24] C. Shang and L. W. Zelazny, "Selective Dissolution Techniques for Mineral Analysis of Soils and Sediments," 2015, pp. 33–80. doi: 10.2136/sssabookser5.5.c3.

[25] B. B. Khaidarov, D. S. Suvorov, D. V. Lysov, G. G. Luchnikova, M. E. Druzhinina, and D. V. Kuznetsov, "Investigation of Mineral Hydraulic Binders Based on the Slag-Cement System Obtained with the Use of Vortex Electromagnetic Homogenization.," *Refractories and Industrial Ceramics*, vol. 62, no. 1, 2021.

[26] P. E. Stutzman, P. Feng, and J. W. Bullard, "Phase Analysis of Portland Cement by Combined Quantitative X-Ray Powder Diffraction and Scanning Electron Microscopy," *J Res Natl Inst Stand Technol*, vol. 121, p. 47, Apr. 2016, doi: 10.6028/jres.121.004.

[27] Y. T. Zhao, X. Chen, L. C. Lu, Y. B. Huang, and J. Zhang, "Determination of Tricalcium Silicates Crystal Forms in Belite-Barium Calcium Sulphoaluminate Cement," *Applied Mechanics and Materials*, vol. 541–542, pp. 204–208, Mar. 2014, doi: 10.4028/www.scientific.net/AMM.541-542.204.

[28] "https://www.mindat.org/."

[29] R. Snellings, M. De Schepper, K. De Buysser, I. Van Driessche, and N. De Belie, "Clinkering Reactions During Firing of Recyclable Concrete," *Journal of the American Ceramic Society*, vol. 95, no. 5, pp. 1741–1749, May 2012, doi: 10.1111/j.1551-2916.2012.05168.x.

[30] Y. Ji, K. Yang, Z. Sun, and W. Zhou, "The Depth Analysis of Different Form Sulfur in the Cement Minerals by Three-Step Extraction Method," *Adv Mat Res*, vol. 524–527, pp. 919–923, May 2012, doi: 10.4028/www.scientific.net/AMR.524-527.919.

[31] Z. F. Li, S. C. Li, Z. H. Zhou, and R. T. Liu, "Study on Extraction of Mesophase from Steel Slag by Potassium Hydroxide-Sucrose (KOSH) Solution Method," *Adv Mat Res*, vol. 1015, pp. 492–496, Aug. 2014, doi: 10.4028/www.scientific.net/AMR.1015.492.

[32] J. J. Thomas, A. J. Allen, and H. M. Jennings, "Structural Changes to the Calcium–Silicate–Hydrate Gel Phase of Hydrated Cement with Age, Drying, and Resaturation," *Journal of the American Ceramic Society*, vol. 91, no. 10, pp. 3362–3369, Oct. 2008, doi: 10.1111/j.1551-2916.2008.02636.x.

[33] C. K. Yip and J. S. J. van Deventer, "Microanalysis of calcium silicate hydrate gel formed within a geopolymeric binder," *J Mater Sci*, vol. 38, no. 18, pp. 3851–3860, 2003, doi: 10.1023/A:1025904905176.

[34] V. K. Peterson and M. C. G. Juenger, "Hydration of Tricalcium Silicate: Effects of $CaCl_2$ and Sucrose on Reaction Kinetics and Product Formation," *Chemistry of Materials*, vol. 18, no. 24, pp. 5798–5804, Nov. 2006, doi: 10.1021/cm061724y.

[35] D. Mombelli, A. Gruttadauria, S. Barella, and C. Mapelli, "The influence of slag tapping method on the efficiency of stabilization treatment of electric arc furnace carbon steel slag (EAF-C)," *Minerals*, vol. 9, no. 11, Nov. 2019, doi: 10.3390/min9110706.

[36] D. Mombelli, G. Dall'Osto, S. Cozzo, and C. Mapelli, "Manipulation of the leaching behaviour of high-alloyed steel EAF slags through the particle size modulation," *IOP Conf Ser Mater Sci Eng*, vol. 1309, no. 1, p. 012008, May 2024, doi: 10.1088/1757-899X/1309/1/012008.

[37] D. Mombelli *et al.*, "Influence of the Recirculation of Various by-products Generated through Electric Arc Furnace Route on EAF Slag Quality," *ISIJ International*, vol. 62, no. 12, p. ISIJINT-2021-571, Dec. 2022, doi: 10.2355/isijinternational.ISIJINT-2021-571.

[38] A. Primavera, L. Pontoni, D. Mombelli, S. Barella, and C. Mapelli, "EAF Slag Treatment for Inert Materials' Production," *Journal of Sustainable Metallurgy*, vol. 2, no. 1, pp. 3–12, Mar. 2016, doi: 10.1007/s40831-015-0028-2.

Journal of Physics: Conference Series **3050** (2025) 012005 doi:10.1088/1742-6596/3050/1/012005

# Smelting of various steel-plant dusts to evaluate recovery of zinc and iron via the Enviroplas™ process

**S Tsebe[1*], S Moloane[1], H Zughbi[2], D Chibwe[2], P Austin[2], D Mchabe[1], M Netshia[1], D Hayman[1], E Matinde[1,3]**

[1] Pyrometallurgy Division, Mintek, Randburg, South Africa
[2] Coke and Ironmaking Technology, BlueScope, Port Kembla, Australia
[3] School of Chemical and Metallurgical Engineering, Wits University, Johannesburg, South Africa

*E-mail: sellot@mintek.co.za

**Abstract.** Steelmaking plants are associated with the generation of large quantities of dust streams from unit processes such as the blast furnace, electric arc furnace, basic oxygen furnace, and argon-oxygen decarburisation furnace. The dust streams are commonly recycled to the smelting process step (in-process recycling). In-process recycling of dust results in the build-up of zinc and other volatile heavy metals such as lead, arsenic, and cadmium, which renders the dust unsuitable for further recycling. The unsuitable dust is either discarded in landfills or temporarily stockpiled pending further processing. If the dusts are stockpiled for extended periods, they tend to pose operational challenges. A cost-effective solution is needed to avoid long term stockpiling of these materials. Mintek developed the Enviroplas™ process to treat such technogenic materials to recover valuable metals and generate a non-hazardous slag by-product. A large pilot smelting campaign was completed in a 1 MW electric arc furnace to demonstrate the recovery of zinc-rich dust and pig iron, as well as an innocuous slag from the processing of various steel-plant dusts via the Enviroplas process. The test work was undertaken as a feasibility study for Australian steelmaker, BlueScope, to evaluate the Enviroplas process as a suitable technology for treating historical stockpiles and new arisings of dust from their steel plant in Port Kembla. The test work demonstrated that zinc-rich dust with 63–75% ZnO, pig iron with 94–98% Fe, and a non-hazardous slag can be produced from the various steel-plant dusts. It was noted that minor operational improvements such as a better feed system design to limit the carryover of feed dust into the gas stream, and efficient control of the bag plant extraction to better define the suction pressure at the gas offtake, can further improve the grade of ZnO in the product dust stream.

## Introduction

The recovery of metals from secondary resources presents numerous benefits over virgin mineral resources. Secondary metal resources refer to technogenic materials from which valuable metals can be recovered [1]. Secondary metal production is resource and energy efficient as it reduces the need for mining and processing of raw ores, generally requires less energy, and produces less carbon dioxide ($CO_2$) emissions than producing the metals from their raw ores [2]. The current study evaluated the pyrometallurgical treatment of metallurgical dusts, not only to mitigate against their environmental impact but also to exploit them as a secondary metal resource.

Metallurgical dusts consists of fine particulates of heterogeneous mixtures of complex oxides from feed and process materials entrained in the off-gas from smelting and refining furnaces [3]. The composition and properties of the dusts vary greatly depending on the furnace type, process conditions, operating procedure, and type of raw materials charged to the furnace [4], [5]. Generally, in-process recycling of dust to the smelter furnace is practised in most integrated ferrous plants, and the common limitation with this practice is the build-up of deleterious and volatile metal compounds of zinc (Zn), lead (Pb), arsenic (As), and cadmium (Cd) [6]. The build-up of these heavy metals renders the dust unsuitable for further recycling, and as a result, the dust is temporarily stockpiled or discarded in landfills. Long term stockpiling or landfilling of ferrous dusts such as blast furnace (BF) dust, electric arc furnace (EAF) dust, and basic oxygen steelmaking (BOS) dust presents serious risks to environmental pollution not only due to the leaching potential of the toxic metals but also the extremely fine nature of the dusts[1], [7]. As such, steel producers are constantly evaluating technologies that can provide a cost-effective solution to not only treat the dust to mitigate against negative environmental impacts, but also to recover valuables metals contained in the dust and generate a non-hazardous by-product in the process.

BlueScope is a global leader in metal coating and painting for building and construction, employing more than 16,500 people at over 160 sites in 15 countries. BlueScope is Australia's largest steel manufacturer and operates an integrated steelworks using the BF-BOS process route at Port Kembla, Australia. BlueScope strongly believes in sustainable manufacturing and is engaged in a range of activities aimed at (i) reducing greenhouse gases emissions, (ii) developing new partnerships aimed at establishing innovative processes for green steel manufacturing, and (iii) improving re-using and recycling of by-products including BOS filter cake (FC) and BF FC which have relatively high contents of iron (Fe) and Zn units. Efficient and innovative re-using of these Fe and Zn rich by-products has been a focus at BlueScope for several years. BlueScope engaged several third parties to test the suitability and select a preferred, cost effective and environmentally friendly technology to recover valuable Fe and Zn units. One of the potential technologies was Mintek's EnviroPlas™ process. Other pyrometallurgical-based technologies that might be suitable to treat BlueScope's BOS FC include Waelz kiln (rotary kiln), rotary hearth furnace, OXYCUP (Cupola), Ausmelt (top submerged lance furnace), coke-packed bed (shaft furnace), electric arc resistance furnace (ESRF), Elkem multipurpose furnace (EMPF), and submerged plasma reactor [8]. Some of the key disadvantages of some of these technologies is the need to agglomerate the feed dust [9], production of direct reduced iron (DRI) instead of pig iron [10], zinc adhesion at furnace top [11], low ZnO grade in product dust and loss of Fe units in slag [12], [13], and higher electrode consumption and more complex operation control [14].

The Enviroplas process is one of the most versatile processes developed by Mintek to treat technogenic solid waste materials from the metallurgical industry. The process is especially suited for problematic solid residues such as steel-plant dusts, stainless steel slags, lead blast furnace slags, and leach residues that contain significant amounts of Zn, Pb, chromium (Cr), nickel (Ni), and other valuable components [15], [16]. The complexity of these technogenic materials and the presence of volatile heavy metal toxins present operational challenges and damage to furnace refractory (if recycled beyond a certain concentration limit) [17]. The Enviroplas process is based on carbothermic smelting of unagglomerated feed using the direct current (DC) electric arc furnace technology, and has been extensively tested and demonstrated at pilot scale using a variety of feed materials [15], [18]. The advantages of the Enviroplas process, for some of the feed materials, are (i) open bath conditions, allowing for the flexibility to process unagglomerated feed materials; (ii) improved kinetics and recoveries achieved in the process, with the ability to achieve test results that are closer to thermodynamic predictions due to the even temperature distribution and improved slag-reductant contact in the bath; and (iii) improved chemistry control and operational flexibility due to the ability to easily control the feed-to-power ratio [15], [16]. Another crucial feature of this process is the ability to generate a non-hazardous by-product slag that can be re-used in other secondary applications (e.g. cement manufacture) or that is safe for disposal in municipal landfill sites [16].

To date, Mintek has completed three phases of test work to test and demonstrate the suitability of the Enviroplas process to BlueScope's various steel-plant dusts. The current paper reports on the test work completed in Phase 3 (large pilot tests). The aim of the project was to advance the engineering knowledge by further demonstrating and validating the process efficiencies of the Enviroplas process in a larger pilot-scale DC arc furnace to increase the confidence level in the feasibility of the process and implementation at a commercial scale. The main objectives of the campaign were (i) to demonstrate the production of metal with a target composition of >96% Fe, <0.5% Si and 2–3% C for each BOS FC stream; (ii) to demonstrate the production of slag with a target composition of 38% $SiO_2$, 25% CaO, 14% MgO, 14% FeO, 7% $Al_2O_3$, and <0.05% of each of PbO and ZnO for each BOS FC stream; (iii) to demonstrate that the process can produce slag that is environmentally safe for disposal as non-hazardous waste for each BOS FC stream; (iv) to demonstrate the collection of dust with a target composition of >65% ZnO, <20% $Fe_2O_3$, and <2% of each of PbO, $Al_2O_3$, MgO, MnO, and $SiO_2$ for each BOS FC stream; and (v) to undertake furnace stack emissions measurements to evaluate the emission rate of combustion gases, particulate matter, metals, volatile organic compounds, and others.

## Experimental method

*Raw materials*

Three types of steel-plant dusts were received from BlueScope, namely Old, New, and Mixed FC (see **Figure 1**(a)-(c)). Old FC was BOS FC that was stockpiled for a long period and had experienced self-sintering. New FC was BOS FC that was not stockpiled for an extended period and had not experienced self-sintering. Lastly, Mixed FC was a mixture of New FC, Old FC, BF FC, BOS metallics, and other secondary particulate materials. Other raw materials were reductant (coke breeze, received from BlueScope), and fluxes (silica and limestone, sourced locally in South Africa). The BOS FCs and coke were received slightly wet with a 10–13% moisture content. These materials were air dried for 3 days to <2% residual moisture, after which they were blended, homogenised, and subsampled for characterisation.

The different streams of BOS FC comprised coarse and fine fractions with a top size of 6.7 mm. The Mixed FC had the most fines fractions while Old FC had the least (see **Figure 1**(d)). Mixed FC had 30% of particles smaller than 300 μm while Old FC had less than 13%. The silica and limestone were coarser with 50% of the particles larger than 5 and 3.5 mm, respectively.

The chemical composition of the BOS FC materials (see **Table 1**) was analysed with Inductively Coupled Plasma Optical Emission Spectroscopy (ICP-OES) for base metals and phosphorus (P). Carbon (C) and sulfur (S) were analysed using a carrier-gas hot extraction technique. The materials predominately contained Fe (49.3-57.6%) with a Zn content between 7 and 14.6%. Other elements such as Ca, Mg, Mn, Si, and Pb were present in minor concentrations (<4%). The chemical compositions of the Old and New FCs were similar but both streams were different to the Mixed FC with respect to Zn and C content. The Mixed FC had lower Zn and higher C content than the other two BOS FCs. Coke breeze had a fixed carbon content of 86.8% and an ash content of 11.7%. Industrial grade silica (96.8% $SiO_2$) and limestone (91.2% $CaCO_3$) were used in the test work.

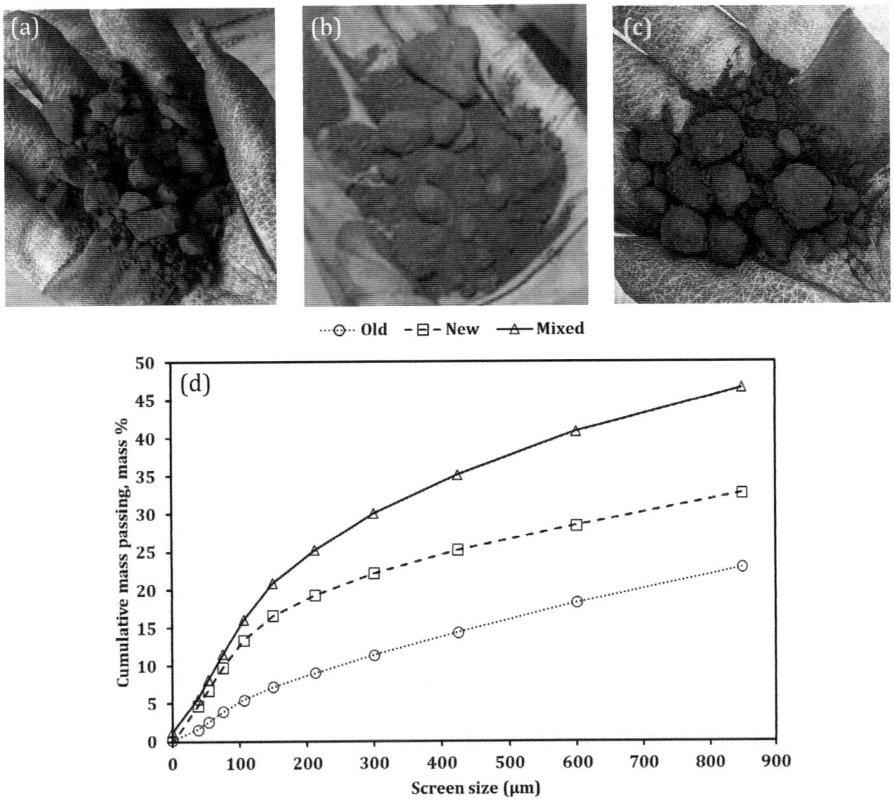


**Figure 1**. Physical appearance of (a)old, (b) New, and (c) Mixed BOS FC, and (d) particle size distribution of the BOS FC materials.

**Table 1**. Elemental chemical composition of BOS FC materials, in mass %.

| BOS FC | Ca | Fe | Mg | Mn | Pb | Si | Zn | C |
|---|---|---|---|---|---|---|---|---|
| Old | 3.58 | 56.8 | 1.32 | 0.76 | 0.28 | 0.58 | 12.0 | 0.76 |
| New | 3.67 | 57.6 | 1.50 | 0.73 | 0.22 | 0.58 | 14.6 | 1.74 |
| Mixed | 3.63 | 49.3 | 1.57 | 0.55 | 0.15 | 0.97 | 7.01 | 6.04 |

Note: Co, Cu, Cr, Ni $<0.05\%$; and Al, S, P $<0.1\%$

The bulk modal phase composition of the BOS FC materials (shown in **Table 2**) was analysed with the Bruker D8 Advance x-ray powder diffractometer. The Old FC was predominantly composed of franklinite (70%) with most of the Fe present as hematite (17%). The New FC comprised mainly wustite (49%) and Fe (25%) with most of the Zn present as zincite (8.6%). Franklinite was present in minor amounts. The Mixed FC contained similar phases but at different proportions of franklinite (31%), wustite (24%), Fe (11%) and hematite (11%) which translated to lower levels of Fe and Zn with increased levels of C. **Figure 2**(a) shows backscattered electron image of Mixed FC showing the major mineral phases such as the franklinite matrix and occurrence of wustite around spherical Fe phases. **Figure 2**(b) shows individual quartz particle and the prevalence of conglomerates of ultra fine particles.

Journal of Physics: Conference Series **3050** (2025) 012005 doi:10.1088/1742-6596/3050/1/012005

**Table 2.** Bulk modal phase composition of BOS FC materials, in mass %.

| Phase | Ideal Formula | Old | New | Mixed |
|---|---|---|---|---|
| Franklinite | $ZnFe_2O_4$ | 70 | 3.7 | 31 |
| Hematite | $Fe_2O_3$ | 17 | - | 11 |
| Wustite | FeO | 4.1 | 49 | 24 |
| Calcite | $CaCO_3$ | 7.3 | 11 | 9 |
| Quartz | $SiO_2$ | 0.1 | 0.1 | 4 |
| Iron | Fe | 1.6 | 25 | 11 |
| Zincite | ZnO | - | 8.6 | 4 |
| Graphite | C | - | - | 5 |
| Hydrotalcite | $Mg_6Al_2(CO_3)(OH)_{16}{\bullet}4(H_2O)$ | - | 2.6 | 1 |
| **Total** | | **100** | **100** | **100** |

Note: Dash (-) means not detected.

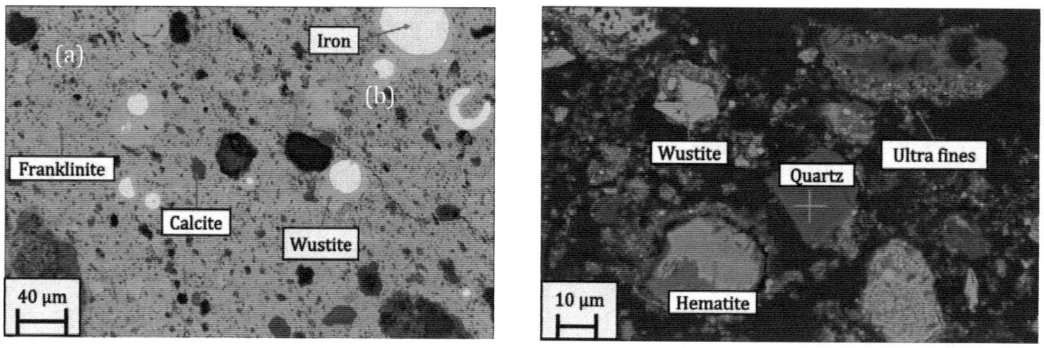


**Figure 2.** Backscattered electron image of Mixed FC showing (a) occurrence of wustite around Fe phases, and (b) individual quartz particle, and ultra fines particles.

*Equipment description*

The pilot plant facility consisted of three main sections, namely the feed system, furnace, and off-gas handling system, as shown in **Figure 3**. The furnace consisted of a 2.5 m cylindrical shell, a domed base, and a conical roof. The installed power supply was rated at 3.2 MVA but the furnace was typically operated at 0.7–1.5 MW. The furnace utilised a single centrally located graphite electrode with a diameter of 200 mm. The shell featured two bi-level tapholes for tapping slag and metal, respectively. The conical roof and upper shell had forced cooling water circuits, while the furnace lower shell utilised film cooling water. The furnace roof (conical and plug) was lined with conventional high-quality alumina castable refractory. The shell (sidewall) was lined with magnesia-chrome refractory bricks and tap blocks. The furnace hearth was lined with two different dry rammables. The bottom part was lined with magnesia-spinel ($MgO-Al_2O_3$) and the top part was lined with a magnesia-lime (MgO-CaO) rammable.

The feed system consisted of batching hoppers in which raw materials were loaded. The raw materials were batched to the surge bins, based on the smelting recipe, before conveyed into a final feed bin. The BOS FC was loaded in a dedicated feed hopper and conveyed to the final feed bin in which it mixed with the other raw materials. The mixed feed materials were charged into

Journal of Physics: Conference Series **3050** (2025) 012005 doi:10.1088/1742-6596/3050/1/012005

the furnace via a screw feeder. Furnace flue gases and carryover dust were extracted through an off-gas port. The gas extraction system included a water-cooled ducting, a reverse-pulse bag filter plant, an extraction fan, and a stack. Since one of the main products of this process was the dust stream, the entire bag filter section along with some areas of the baghouse were thoroughly cleaned and 396 new filter bags were installed before the campaign.

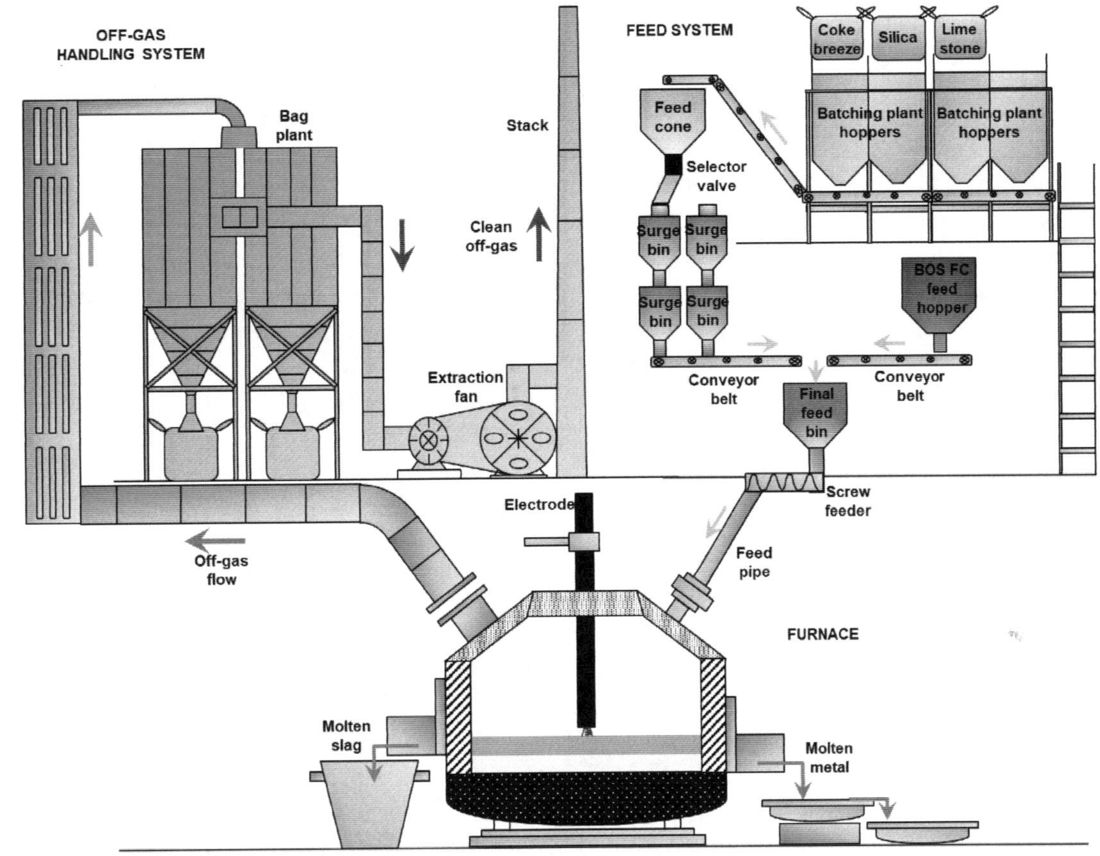


**Figure 3**. General layout of 1 MW DC arc furnace showing the main sections of the pilot plant.

*Operating procedure*

The furnace was loaded with start-up Fe alloy, coke, and anthracite, and commissioned by taking it through a 24–30 h warm-up to slowly preheat the refractory materials. This was followed by slow feeding in the initial stages whereby the furnace was operated with a lower power input. The test work aimed to operate the furnace in a manner that provided the necessary process conditions to promote high recoveries of Zn to the dust stream and Fe to the metal stream. This required the application of an appropriate operating temperature and the addition of suitable amounts of reductant and fluxing reagents. An initial estimate of the required process conditions was estimated with PyroSim™, Mintek's steady state chemical thermodynamic simulation software. The software was used to calculate the mass and energy balance of the process flowsheet to smelt each BOS FC (given a set of inputs). The specific energy requirement (SER) of the process for smelting Old, New, and Mixed FC was estimated at 1.149, 1.142, and 1.104 kWh/kg total feed, respectively. These values, along with the corresponding amounts of reductant and fluxes, were used in the smelting campaign as starting conditions. Process information collected

Journal of Physics: Conference Series **3050** (2025) 012005 doi:10.1088/1742-6596/3050/1/012005

after each batch (e.g. tapping temperatures, product masses, chemical compositions, furnace heat losses) was used to inform decisions about how to adjust the SER, reductant, and flux additions for subsequent batches in order to achieve the desired product yield and quality.

Different power levels were also investigated to assess the performance of the furnace and its effect on the product when the furnace was operated with high feed rates. The furnace's electrical parameters were also adjusted accordingly to ensure the furnace delivered the required energy to the process to obtain the desired feed-to-power balance.

Molten products were tapped from two different tap holes. Slag was tapped into cast iron ladles whereas metal was tapped into refractory lined ladles at the end of each batch. Sampling of the product metal and slag was undertaken in the molten state. A metal spoon was passed through a molten stream of metal or slag during the tapping of the furnace. The molten metal sample was quenched in a drum of stirred water while the molten slag sample was air cooled by pouring it onto a cold metal plate. The dust was sampled from the bulk bags collecting dust from the bag plant. The dust in each bulk bag was first thoroughly mixed to homogenise it. Five subsamples were taken using a 500 mm long pipe at different points. All the subsamples were combined and blended before a representative subsample was withdrawn using a rotary riffle splitter. The product samples were analysed with ICP-OES for base metals and P. Appropriate preparation methods were adopted for the respective product matrices. Carbon and S were analysed using the carrier-gas hot extraction technique.

## Results and discussion

*Campaign overview*

The campaign processed 95 t of BOS FC materials over 75 batches. This was associated with an overall addition of 19% (per kg BOS FC) coke, 6.6% silica, and 3.4% limestone. Four BOS FC smelting conditions were evaluated. That is, smelting of (i) 29 t of Old FC, (ii) 25 t of New FC, (iii) 7 t of Blend FC which was composed of 3.1 t of Old FC (44%) and 3.9 t of New FC (56%), and (iv) 32 t of Mixed FC. The batch feed masses for each tested condition are presented in **Table 3**. The total raw materials fed and products generated for each BOS FC condition are given in **Figure 4**. The overall campaign obtained a metal yield of 38% of the total BOS FC materials processed, and collection of dust equivalent to 14% of the total mass of materials fed to the furnace.

**Table 3**. Batch feed recipe for each BOS FC smelting condition.

| BOS FC | Tap | Batch BOS FC Mass, kg | Mass fed^, mass % Coke | Silica | Limestone |
|---|---|---|---|---|---|
| Old | 1-7 | 1000 | 22% | 9% | 0% |
| | 8-13 | 1000 | 25% | 8% | 0% |
| | 15-16 | 1000 | 22% | 6% | 0% |
| | 17-21 | 1000 | 23% | 6% | 5% |
| New | 22-27 | 1000 | 23% | 6% | 5% |
| | 28-33 | 1000 | 21% | 6% | 5% |
| | 34-39 | 1000 | 19% | 6% | 5% |
| | 40-42 | 1000 | 17% | 6% | 5% |
| Blend | 43-49 | 1000 | 17% | 6% | 5% |
| Mixed | 50-51 | 1000 | 13% | 6% | 4% |
| | 52-75 | 1000 | 15% | 6% | 4% |

^ Mass fed per kg of BOS FC

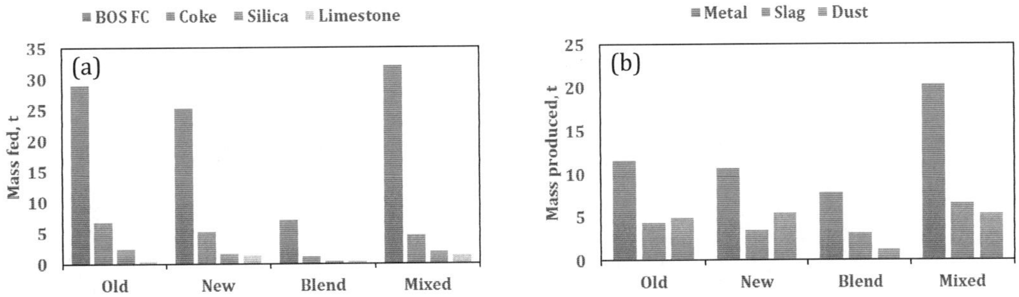


**Figure 4**. Total mass of raw materials processed, and products generated for each BOS FC condition.

The average furnace operating parameters and obtained tapping temperatures are presented in **Table 4**. The smelting of Old and Mixed FC was evaluated under two different power inputs. The Old FC was smelted with a furnace power of 750 and 800 kW while the Mixed FC was smelted with power inputs of 800 and 950 kW. The New and Blend FC were smelted with a furnace power of 800 kW. The furnace was operated with voltages of 260–320 V to ensure an open arc configuration and open bath condition was maintained. The furnace was operated with an SER and heat loss setpoint of 1.1–1.2 kWh/ kg and 300–400 kW, respectively. The overall campaign measured SER and heat loss was 1.18 kWh/ kg and 441 kW, respectively. The product slag was tapped at average temperatures of 1341–1560°C and metal was tapped at 1323–1500°C.

**Table 4**. Average furnace operating parameters and obtained tapping temperatures for each BOS FC smelting condition.

| BOS FC | Power, kW | Current, A | Voltage, V | SER*, kWh/ kg | Heat loss, kW | Metal temp, °C | Slag temp, °C |
|---|---|---|---|---|---|---|---|
| Old (LP^) | 750 | 2885 | 260 | 1.2 | 300 | 1385 | 1460 |
| Old (HP#) | 800 | 2996 | 267 | 1.2 | 400 | 1500 | 1560 |
| New | 800 | 2996 | 267 | 1.2 | 400 | 1403 | 1474 |
| Blend | 800 | 2996 | 267 | 1.2 | 400 | 1430 | 1484 |
| Mixed (LP^) | 800 | 2996 | 267 | 1.1 | 400 | 1491 | 1394 |
| Mixed (HP#) | 950 | 2969 | 320 | 1.1 | 400 | 1323 | 1341 |

^ LP means lower power. # HP means higher power. * SER means specific energy requirement

The campaign achieved an overall recovery of 89% Zn to the dust and a recovery of 84% Fe to the metal. The accuracy of the recovery calculation is dependent on complete accountability of all material masses entering, leave, and accumulating in the system. In a pilot plant facility, it is often challenging to accurately recover all the accumulated mass in the ducting of the off-gas handling system. Thus, negatively affecting the accountability of some of the elements. The deportment formulation provides a better assessment of the preferential partitioning of the major elements to the respective product phases. The elemental deportments to the respective solid product streams for each BOS FC smelting condition are presented in **Figure 5**. The results clearly show that almost 100% of Zn reported to the dust stream for all BOS FCs smelting conditions. A similar behaviour was observed for Pb (and possibly for other volatile elements such as As and Cd). The deportment of Fe to the metal stream for Old, New, Blend, and Mixed FC was 92, 92, 95, and 88%, respectively. It was observed that a considerable amount (31%) of Mn reported to the metal

Journal of Physics: Conference Series **3050** (2025) 012005 doi:10.1088/1742-6596/3050/1/012005

stream and 43% of Mg reported to the dust stream during the smelting of New FC. Other elements that generally form stable metal oxides predominately reported to the slag stream for all BOS FC smelting conditions.

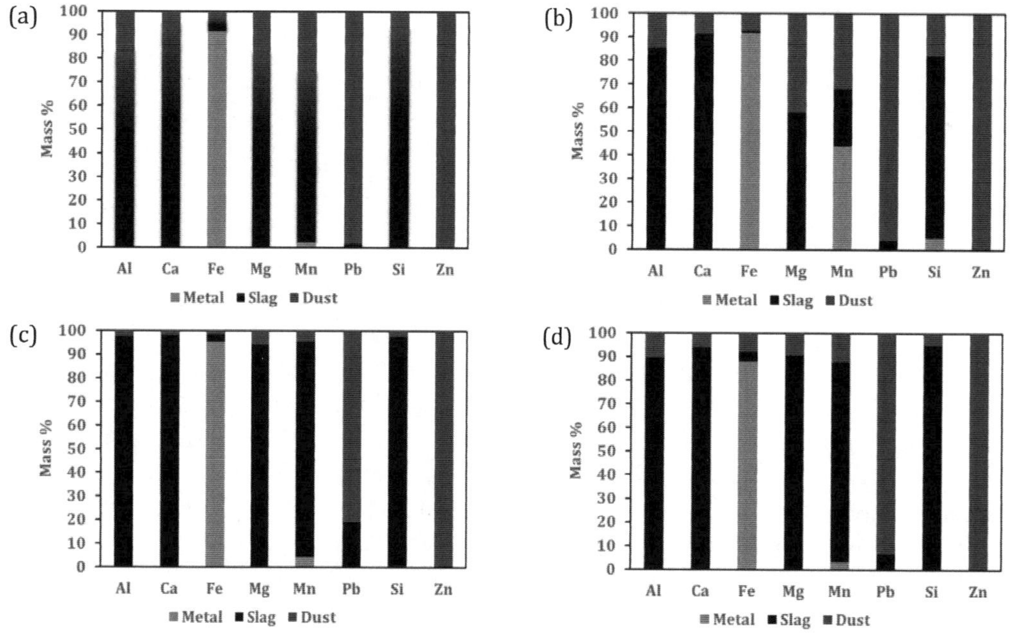


**Figure 5**. Elemental deportment from (a) Old, (b) New, (c) Blend, and (d) Mixed BOS FC smelting conditions.

*Metal*

The average chemical composition of the metal produced under each BOS FC smelting condition is given in **Table 5**. The profile of the composition of Fe, C, and Si in metal is presented in **Figure 6**, and that of S, P, Cr, and Mn is presented in **Figure 7**. Production of metal with the target composition of >96% Fe, <0.5% Si and 2–3% C was successfully demonstrated during the smelting of Old, Blend, and Mixed FC. The target metal composition was not demonstrated during the smelting of New FC. This condition was characterized by the highest C content (>3%) in metal. This high C content was associated with higher occurrence of other minor impurities such as Cr, Mn, and Si. The C and the other minor impurities remained consistently high in the metal despite several attempts to drop the C addition. The condition consisted of 21 batches and the C addition was dropped from 20 to 14.8% by making approximately 2% change every 6 to 7 batches. However, this did not help decrease the level of impurities in the metal. The Fe grade dropped to a low of 93.6% at which point the C content was at a high of 3.98%.

**Table 5**. Weighted average chemical composition of the metal tapped under each BOS FC smelting condition, in mass %.

| BOS FC | Cr | Fe | Mn | Si | C | S | P |
|---|---|---|---|---|---|---|---|
| **Target** | **-** | **>96** | **-** | **<0.50** | **>2** | **-** | **-** |
| Old | 0.17 | 97.9 | 0.03 | 0.03 | 1.46 | 0.18 | 0.17 |
| New | 0.51 | 94.8 | 0.54 | 0.38 | 3.52 | 0.08 | 0.15 |
| Blend | 0.15 | 97.4 | 0.05 | 0.03 | 2.27 | 0.12 | 0.01 |
| Mixed | 0.14 | 97.2 | 0.05 | 0.01 | 2.31 | 0.20 | 0.05 |

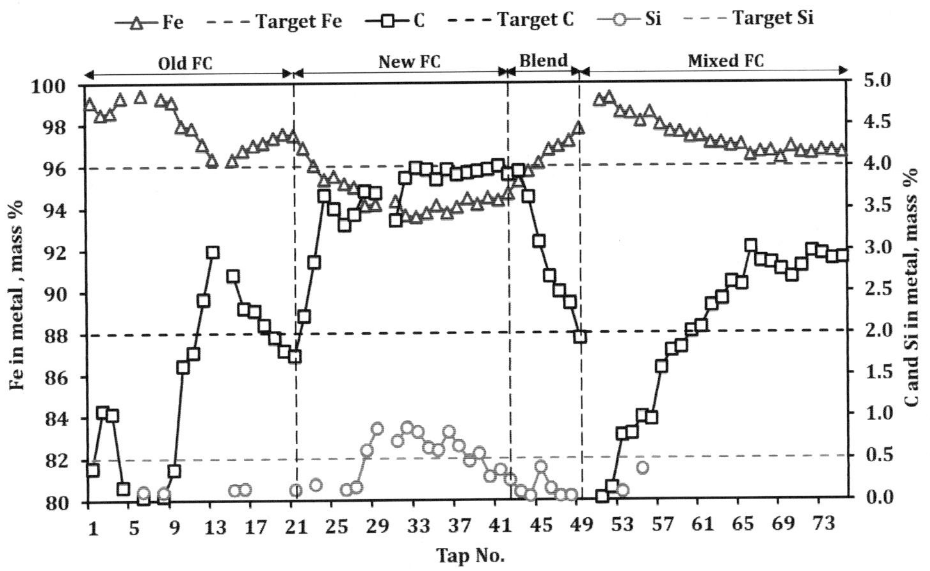


**Figure 6**. Composition of Fe, C, and Si in metal for each BOS FC smelting condition.

There was a clear indication that the New FC behaved differently from the other BOS FC streams during smelting. At the same C addition (e.g. 20%), Old FC condition produced metal with C between 1.5 and 2.5% but immediately switching to New FC, the C in metals produced abruptly increased to above 3.5%. This difference in the smelting behaviour of New FC was largely due to its phase chemical composition. New FC comprised 48.8% wustite, 25% Fe and only 3.7% franklinite while Old FC was predominantly composed of franklinite (69.7%) with most of the Fe present as hematite (17.2%). Old FC contained much higher amounts of $O_2$ in its oxide phases leading to higher C requirements during smelting while New FC had less C requirements due to a lower content of oxygen atoms in its oxide phases as well as a higher content of metallic Fe. This explains the observed higher concentrations of impurities in metal during smelting of New FC as there was more C available to react with metal oxides of impurity elements such as $SiO_2$, $Cr_2O_3$, and MnO. The Mixed FC behaved similarly to Old FC as it also contained higher amounts of franklinite and hematite than New FC.

It was also observed that the S in metal dropped as low as 0.03% during the smelting of New FC. This signified that the deportment of S to the metal was inversely proportional to C content in the metal. The higher the concentration of C in metal, the lower the concentration of S in metal. That is, the removal of S from metal was favoured by highly reducing conditions.

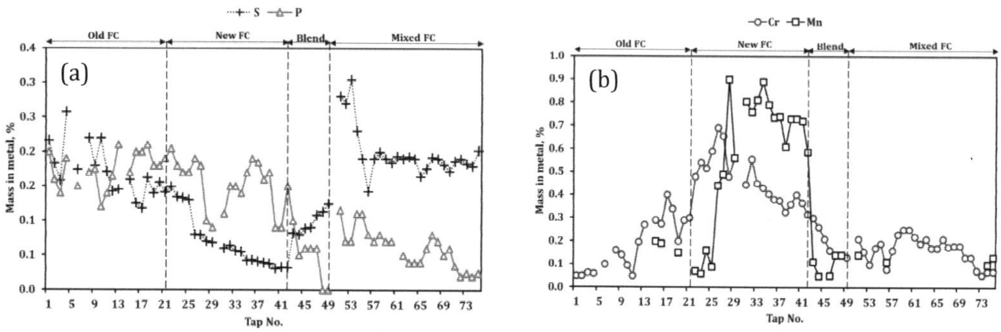


**Figure 7**. Composition of (a) S and P, and (b) Cr and Mn in metal for each BOS FC smelting condition.

*Slag*

The weighted average chemical composition of the slag tapped under each BOS FC smelting condition is given in **Table 6**. The compositional profiles of the slags are presented in **Figure 8**. The slag compositions for the different conditions varied significantly. This was largely due to the recipe adjustments that were implemented in each smelting condition. The Mixed FC condition had the most consistent slag compositions with narrower compositional variation ranges of the oxides as the smelting recipe was not changed during this condition. There was a considerable increase in CaO during the New FC condition. The increase in CaO was a result of limestone addition which was introduced from Tap 17. Naturally, this diluted other oxides with the obvious dilution observed for $SiO_2$. The surprising response was the notable increase in the concentration of $Al_2O_3$ from 5 to 10% in Tap 16. This higher level of $Al_2O_3$ was sustained in the slag between 8 and 10% during the New FC condition and eventually dropped back to around 5% during the Mixed FC condition. The CaO concentration also dropped to between 25 and 30% during the Mixed FC condition as limestone addition was decreased to 4% from an addition of 5% in all previous conditions.

As a result of the significant variation in slag composition throughout the campaign, the test work was unable to demonstrate the production of slag, during the smelting of any of the BOS FC, whereby the target compositions of 38% $SiO_2$, 25% CaO, 14% MgO, 14% FeO, and 7% $Al_2O_3$, were all met. The weighted average composition of slags from the Old FC condition was close with $SiO_2$ (36.6%) and $Al_2O_3$ (6.24%). Slag from New FC was also close with $SiO_2$ (38.9%) and $Al_2O_3$ (8.95%). For this condition, CaO overshot the target at 34.8% while FeO significantly undershot the target at 3.23%. These results were due to the combined effect of the sudden availability of excess C owing to the lower C requirement of New FC and additions of limestone at 5%. Slag from the Blend FC condition was close to target also with $SiO_2$ (37.1%) and $Al_2O_3$ (6.99%). The slag from the Mixed FC condition was close with $SiO_2$ (37.7%), $Al_2O_3$ (5.45%), and CaO (27.7%). CaO composition was closer to target in this condition mostly because the limestone addition was dropped to 4%. Although the composition of MgO was slightly below target, it was the most consistent in all conditions at approximately 11% while the FeO content had the widest variation (3.23–18.4%).

The expected corresponding increase in S content in the slag was observed during the New FC condition due to preferential deportment of S from metal to slag under highly reducing conditions induced during this condition.

Journal of Physics: Conference Series **3050** (2025) 012005 doi:10.1088/1742-6596/3050/1/012005

**Table 6**. Weighted average chemical composition of the slag tapped under each BOS FC smelting condition, in mass %.

| BOS FC | $Al_2O_3$ | CaO | $Cr_2O_3$ | FeO | MgO | MnO | PbO | $SiO_2$ | ZnO | Total |
|---|---|---|---|---|---|---|---|---|---|---|
| **Target** | **7.0** | **25** | **-** | **14** | **14** | **2.0** | **<0.05** | **38** | **<0.05** | |
| Old | 6.24 | 15.9 | 1.28 | 18.4 | 11.8 | 3.23 | 0.04 | 36.6 | 0.06 | 94.2 |
| New | 8.95 | 34.8 | 0.10 | 3.23 | 11.3 | 1.18 | 0.05 | 38.9 | 0.16 | 99.5 |
| Blend | 6.99 | 30.3 | 0.66 | 9.67 | 10.9 | 3.34 | 0.09 | 37.1 | 0.14 | 100 |
| Mixed | 5.45 | 27.7 | 0.69 | 10.2 | 11.9 | 2.73 | 0.05 | 37.7 | 0.10 | 97.6 |

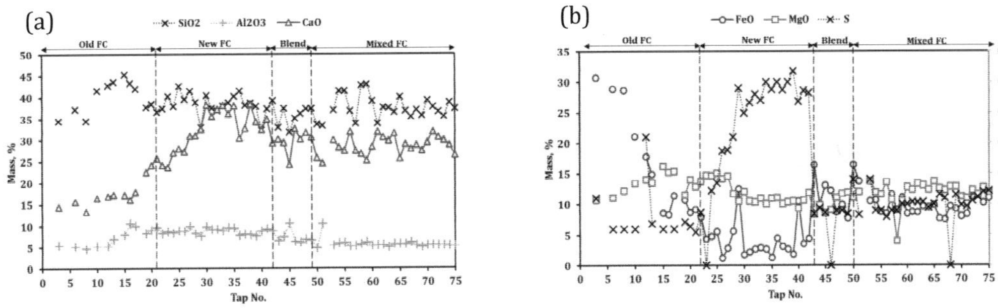


**Figure 8**. Composition of (a) $SiO_2$, $Al_2O_3$, and CaO, and (b) FeO, MgO, and S for each BOS FC smelting condition.

The slag basicity, CaO/SiO2 ratio, slag tapping and liquidus temperatures, and slag viscosity are presented **Figure 9**. The slag basicity (expressed as $(CaO+MgO)/SiO_2$) was mostly above the target of 1.0 throughout the campaign. This was favourable as it minimised chemical attack on the refractory walls and prevented excessive dissolution of MgO into slag. This was also supported by the consistent MgO concentration in slag throughout the different smelting conditions. The liquidus temperatures and the viscosity of the molten slags at the slag tapping temperatures were estimated using Factsage 8.1. Viscosities were calculated with the Viscosity module and the liquidus temperatures were estimated with the Equilib module. The calculations were based on the composition of the major components of the slag ($Al_2O_3$, CaO, FeO, MgO, MnO and $SiO_2$) and the FToxid database was selected. Generally, all estimated slag liquidus temperatures were below the slag tapping temperatures with the exception of slag from Taps 58–63 and Taps 70–74. The profiles indicated an expected correlation between slag tapping temperature and viscosity. Slags with lower tapping temperatures were associated with relatively higher viscosities.

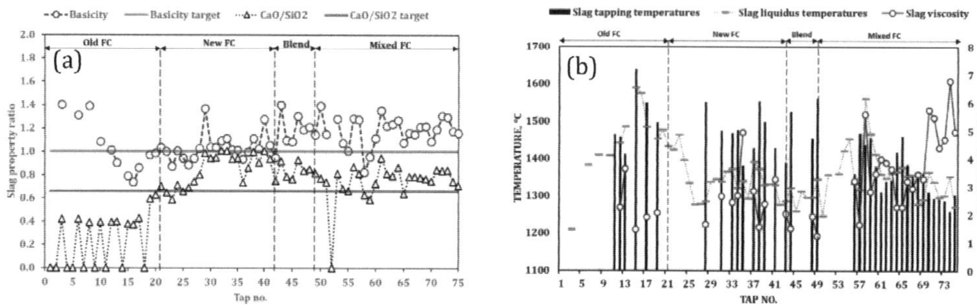


**Figure 9**. (a) Slag basicity and $CaO/SiO_2$ ratio, and (b) slag tapping and liquidus temperature, and slag viscosity for each BOS FC smelting condition.

Another important aspect of the test work was to demonstrate that the process could produce a slag that was environmentally safe for disposal as non-hazardous waste for each BOS FC stream. Selected slag samples from each BOS FC stream were submitted for toxicity characteristic leaching procedure (TCLP) tests [19] to determine the leaching rate of toxic heavy elements such as Pb, As, Cd, and Hg. **Table 7** presents the results of the TCLP tests, and it is shown that all the elements of interest have leaching rates below 0.1 ppm. According to the New South Wales Environment Protection Authority (NSW EPA) waste classification guidelines [20], this classifies the tested slag samples as non-hazardous waste. Therefore, the test work demonstrated that the process can produce slags that are environmentally safe for disposal for each BOS FC stream.

**Table 7**. TCLP results and limits for slag, in parts per million (ppm).

| Sample | BOS FC | | | NSW EPA |
|---|---|---|---|---|
| | Old^ | New# | Mixed* | limits, TCLP |
| Pb | <0.1 | <0.1 | <0.1 | 20 |
| As | <0.1 | <0.1 | <0.1 | 20 |
| Cd | <0.1 | <0.1 | <0.1 | 4 |
| Hg | <0.1 | <0.1 | <0.1 | 0.8 |
| Mo | <0.1 | <0.1 | <0.1 | 20 |
| Ni | 0.1 | <0.1 | <0.1 | 8 |

^ Tap 19 slag, # Tap 41 slag, and * Tap 68 slag were selected for TCLP tests

*Dust*

The weighted average chemical composition of dust collected under each BOS FC smelting condition is given in **Table 8**. The compositional profiles of major and minor components in the dust are illustrated in **Figure 10** and **Figure 11**, respectively. The composition of the dust varied significantly with the different smelting conditions, and within each condition. The wide variations were observed with almost all the major components in the dust. The primary component, ZnO, varied between 50% and 86%. The primary contaminant to the dust was $Fe_2O_3$ with secondary noticeable contributions from $SiO_2$ and MgO especially during the New FC condition. The ZnO grade of the dust was inversely proportional to the concentration of these contaminants in the dust. The higher the concentration of contaminants, the lower the ZnO grade in the dust. Recovery of these contaminants to the dust stream occurred predominantly by carryover of feed dust to the off-gas stream. Overall, dust collected during the Old FC condition had the highest weighted average ZnO grade of 75.4% and dust from the Mixed FC condition had the lowest ZnO grade of 62.4%. This finding correlated very well with the PSD of the respective BOS FC streams. That is, Old FC had the least quantity of fines which minimised feed carryover while Mixed FC comprised the most quantity of fines which increased feed carryover. The New and Mixed FC conditions experienced high feed carryover with the $Fe_2O_3$ averaging 20.6 and 30.3%, respectively. As such, only Old and Blend FC conditions demonstrated collection of dust with a target composition of >65% ZnO, <20% $Fe_2O_3$, and <2% of PbO, $Al_2O_3$, MgO, MnO, and $SiO_2$. In both conditions, the highest grades achieved had more than 80% ZnO in dust.

**Table 8.** Weighted average chemical composition of the dust collected under each BOS FC smelting condition, in mass %.

| BOS FC | $Al_2O_3$ | $Cr_2O_3$ | $Fe_2O_3$ | MgO | PbO | $SiO_2$ | ZnO | S | Total |
|---|---|---|---|---|---|---|---|---|---|
| Target | <2 | <2 | <20 | <2 | <2 | <2 | >65 | <2 | |
| Old | 1.12 | 1.43 | 12.1 | 2.18 | 1.80 | 2.27 | 75.4 | 0.18 | 98.8 |
| New | 0.95 | 0.58 | 20.6 | 5.02 | 0.80 | 5.54 | 63.8 | 0.18 | 101 |
| Blend | 0.44 | 0.18 | 14.9 | 1.67 | 0.94 | 2.45 | 70.3 | 0.14 | 93.2 |
| Mixed | 0.80 | 0.17 | 30.3 | 1.50 | 0.96 | 2.50 | 62.4 | 0.20 | 102 |

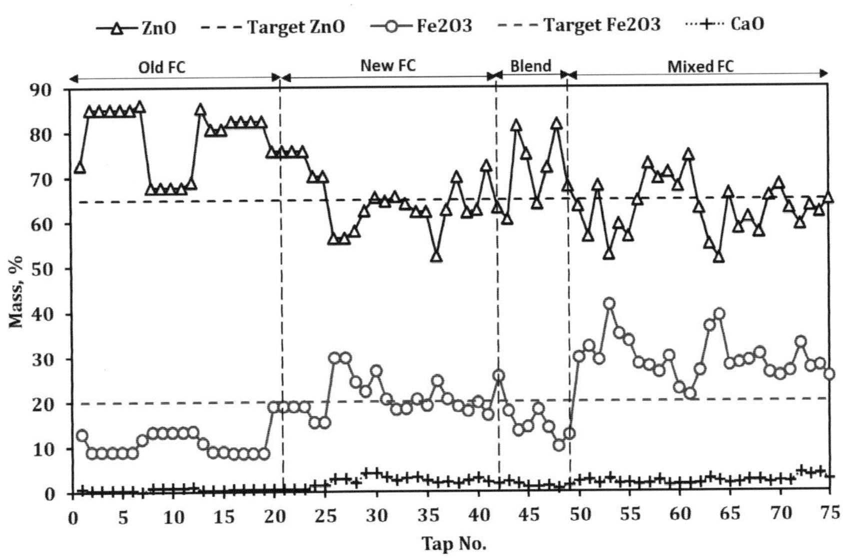


**Figure 10.** Composition of ZnO, $Fe_2O_3$, and CaO in the dust for each BOS FC smelting condition.

The New FC condition experienced a well-pronounced spike in minor components such as $SiO_2$, MgO, $Al_2O_3$, MnO, and $Cr_2O_3$ (see **Figure 11**(a)).The same minor elements also spiked in concentration during Old FC condition from dust collected between Tap 8 and 12. The observed abrupt increase in concentration was associated with periods of high C addition and/or availability of excess C. Under highly reducing conditions and high operating temperatures (>1400°C), some stable oxides can be reduced to form volatile metals (e.g. $MgO(s) + C(s) = Mg(g) + CO(g)$) or volatile gases (e.g. $SiO_2(s) + C(s) = SiO(g) + CO(g)$). It is these reactions that are believed to have led to the high deportment of elements such as Si, Mg, and Mn to the dust stream during periods of high C addition. This was further supported by the observed gradual decrease in concentration of these oxides in the dust with decreased C additions applied during the New FC condition.

There was also a considerable increase in the concentration of $SiO_2$ in dust (2–4.2%) during the Mixed FC condition. A size-by analysis of $SiO_2$ in the different BOS FC showed the occurrence of a high concentration of $SiO_2$ in the −0.2 mm size class of Mixed FC. It is believed that carryover of these fines contributed to the observed increase in the concentration of $SiO_2$ in the dust during the Mixed FC condition.

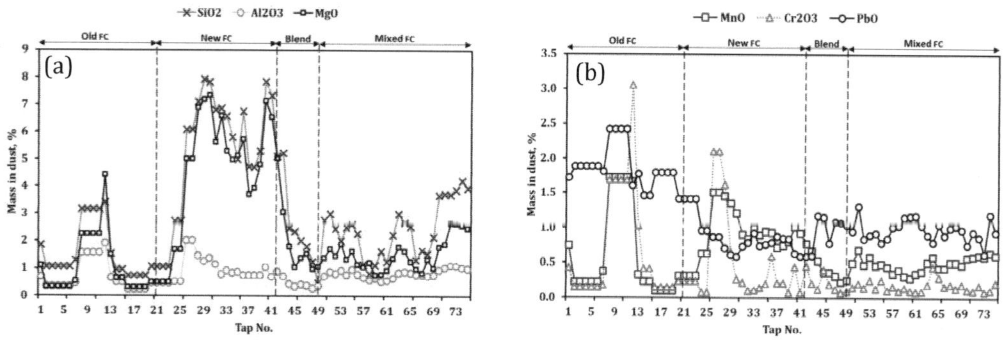


**Figure 11**. Composition of (a) $SiO_2$, $Al_2O_3$, and MgO, and (b) MnO, $Cr_2O_3$, and PbO in the dust for each BOS FC smelting condition.

*Effect of location of feed port*

Findings from the Phase 2 test work determined that the quality of the dust was severely affected by feed carryover. One of the recommendations from the test work was to change the position of the feed port. In the Phase 2 test work, the feed port was located relatively close to the off-gas port, as shown in **Figure 12**(a). It was recommended to locate the feed port on the opposite side of the off-gas port (at 180°) with the electrode positioned in between the two ports (as illustrated in **Figure 12**(b)). This design of the furnace roof and the location of ports was implemented in the current test work.

The Mixed FC condition can be used to assess whether the location of the feed port, in relation to the off-gas port, had any effect on the quality of the dust. The Mixed FC condition was ideal as it applied a low C addition, had no recipe changes, and the Mixed FC material had the most content of fines. The results showed that the dust quality was still negatively impacted by feed carryover. The Mixed FC condition had the highest concentration of $Fe_2O_3$ in the dust. Consequently, this led to the lowest average ZnO grade (62.4%) in the dust. Thus, locating the feed port on the opposite side of the off-gas port, with the electrode position in between the ports, did not mitigate feed carryover nor help improve the ZnO grade of the dust. However, this should be further considered in conjunction with the intensity of the off-gas suction as discussed in the next section.

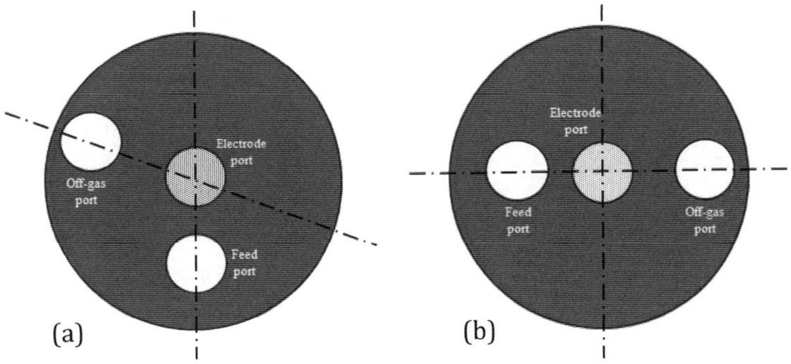


**Figure 12**. Schematic showing position of the feed port in relation to off-gas port on (a) Phase 2 furnace, and (b) Phase 3 furnace.

*Off-gas suction intensity*

Another crucial process parameter that played a role in the management of dust quality was the intensity of the off-gas suction from the baghouse extraction fan. This parameter not only influenced the quality of the dust due to feed carryover but also impacted (i) the rate of off-gas port blockage; (ii) control of fugitive emissions through electrode port and feed pipe; and (iii) the off-gas inlet temperature into the baghouse. It was determined that during the operation of the furnace, the suction intensity of the bag plant extraction fan had to be adjusted to an optimal level to improve the efficiency of operation of the furnace and control of fugitive emissions. A high suction intensity increased feed carryover and off-gas inlet temperature into the baghouse. A low suction intensity increased fugitive emissions through the feed pipe and dust losses in the off-gas ducting system. It was also observed that this optimal level might change during a batch due to (i) gradual build-up of dust in the filter bags (which reduces suction pressure); (ii) gradual build-up of accretions in off-gas port (which reduces open port area); and (iii) gradual increase of off-gas inlet temperature into baghouse (which increases the risk of damaging baghouse filter bags)

As a result, the design, capacity, and optimal operation of the off-gas treatment plant form an integral part of the efficient operation of this process. Since the quality of dust, in respect of ZnO grade, determines the economic viability of this processing technology, it is imperative that the off-gas treatment plant is adequately designed and sufficiently instrumented to enhance control and effective management of the process.

*Energy requirement and thermal efficiency*

The SER of the overall process was calculated in two ways from the data generated during the campaign. Firstly, the SER was calculated by plotting the average power vs. average tap feed rate for the entire campaign and this can be seen in **Figure 13**(a). The slope of the line estimates the SER of the process (independent of heat losses), and the heat loss value is determined by the y-intercept. From this method of calculation, the furnace heat losses are underestimated (317 kW) and the SER is overestimated (1.44 kWh/kg). The $R^2$ value is too low (<0.9) indicating that the trend line does not fit the data accurately. This is normally the case when there are several outlying data points due to unstable or abnormal operating conditions. In the second method, the SER was calculated from the total energy delivered to the furnace divided by the total mass fed to the furnace (1.18 kWh/kg). This SER is more credible and similar to those calculated from the preliminary simulations (1.104–1.149 kWh/kg), and the SER obtained for the overall campaign from the Phase 2 test work (1.10 kWh)/kg. The average heat losses from the furnace for the entire campaign was 441 kW.

The heat losses to the environment used in the energy balance relate to the thermal efficiency of the equipment. In general, the thermal efficiency of a furnace increases with size i.e. higher power, higher thermal efficiency. The thermal efficiency of a small-scale furnace is expected to be much lower than that of a commercial or large-scale pilot plant facility. The furnace is fitted with a water-cooled jacket covering most of the furnace sidewalls and a water-cooled roof. A plot of the furnace power vs. thermal efficiency for the period of feeding Mixed FC condition is shown in **Figure 13**(b). The average thermal efficiency during this period was 44%.

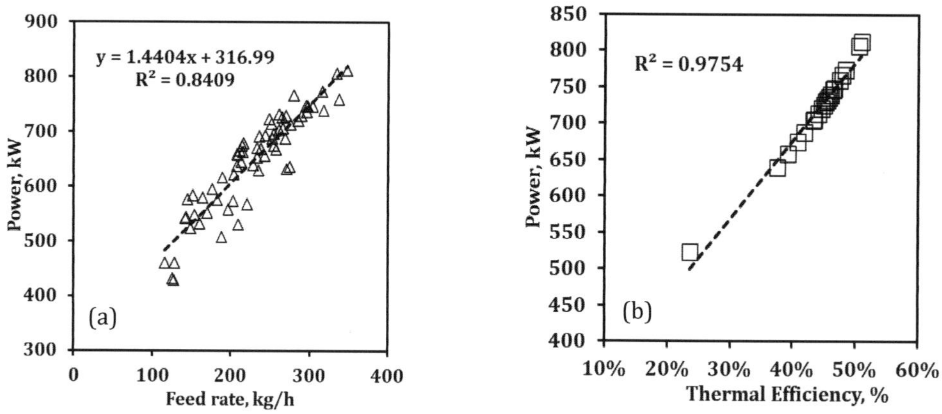


**Figure 13**. Plot of (a) furnace power against feed rate, and (b) furnace power against thermal efficiency.

*Electrode consumption*
The overall electrode consumption measured during the campaign was 2.36 kg/MWh. For comparison, the electrode consumptions measured for other processes conducted in the same pilot facility were 0.78 kg/MWh for chromite smelting, 1.4 kg/MWh for ilmenite smelting, and 2.3 kg/MWh for DRI smelting. As such, the electrode consumption was similar to that obtained during DRI smelting. The reason the electrode consumption from chromite and ilmenite smelting were lower than the consumption obtained in the current test work was mainly due to the difference in the volume off-gas generated. The current test work had much higher volumes of gaseous species generated per unit mass treated. This required higher extraction intensity, which led to increased electrode consumption. There was also more contact of electrode with volatilised species, which may also react with the electrode and increase the consumption rate.

*Furnace emission measurements*
The measurements of furnace emissions were undertaken in accordance with internationally accepted reference methods, which comply with the National Environmental Management: Air Quality Act of 2004 [21]. Two sets of measurements were undertaken to evaluate emissions during the smelting of Blend FC and Mixed FC.

For the Blend FC condition, the results indicated that the process released insignificant volumes of $SO_2$ gas. CO emissions had an average concentration of 25.97 mg/Nm$^3$ which is

Journal of Physics: Conference Series **3050** (2025) 012005 doi:10.1088/1742-6596/3050/1/012005

equivalent to emissions of 5215 kg/year (assuming plant runs 24 hrs/day, 365 day/year). Particulate matter emissions were measured at 2.78 mg/Nm$^3$ (283 kg/year), TVOC at 6.41 mg/Nm$^3$ (1301 kg/year), and total metals at 0.07 mg/Nm$^3$ (13.5 kg/year). For the Mixed FC condition, the results indicated that the process also released insignificant volumes of $SO_2$ gas. CO emissions had an average concentration of 42.5 mg/Nm$^3$ which is equivalent to emissions of 6075 kg/year. Particulate matter emissions were measured at 2.07 mg/Nm$^3$ (193 kg/year), TVOC at 8.6 mg/Nm$^3$ (1272 kg/year), and total metals at 0.07 mg/Nm$^3$ (9.6 kg/year). It should be noted that all the above measured emissions are significantly lower than those specified in the NSW Protection of the Environment Operations Act of 1997 [22]. For example, the emissions limit for TVOC is 40 mg/Nm$^3$ and total Cd and Hg is 0.4 mg/Nm$^3$.

*Refractory performance*
An image showing a top view of the inside of the furnace (with roof removed) after the campaign, is depicted in **Figure 14**. This image shows how the furnace sidewall refractory was sectioned to undertake measurements of the refractory wear. **Figure 15** presents refractory wear profiles of the four sections outlined. The material remaining in the furnace after the furnace was drained occupied a height of 180 mm from the hearth. Measurements were taken from this height, which was equivalent to the height of the metal taphole. As expected, severe refractory wear was observed on areas of the sidewall that were around the taphole. Sidewall refractory around the tapholes commonly experience the most wear due to oxygen lancing during tapping operations. This is shown by side A of section AB (metal taphole) and side G of section GH (slag taphole). Generally, minimal wear was observed on the areas of the sidewall that were not located close to the tapholes. That is, side B of section AB, side H of section GH, and side F of section EF.

However, some wear was observed on areas of the sidewall that were not located close to the tapholes. This was especially the case with side C of section CD and to a lesser extent, side E of section EF. It was surprising that despite their location relative to the tapholes, they also experienced considerable wear. It is suspected that this was due to inadequate cooling around the area of the shell where these bricks were located. Inadequate cooling of the shell prevented formation of a freeze lining to protect the refractory bricks from chemical erosion by the slag.

Overall, it can be concluded that the mag-chrome refractory bricks performed satisfactorily with respect to providing adequate containment of the molten products under the operating conditions tested. However, it is noted that formation of freeze lining is crucial for obtaining satisfactory performance and an extended service life of the refractory bricks under the prevailing slag systems tested.

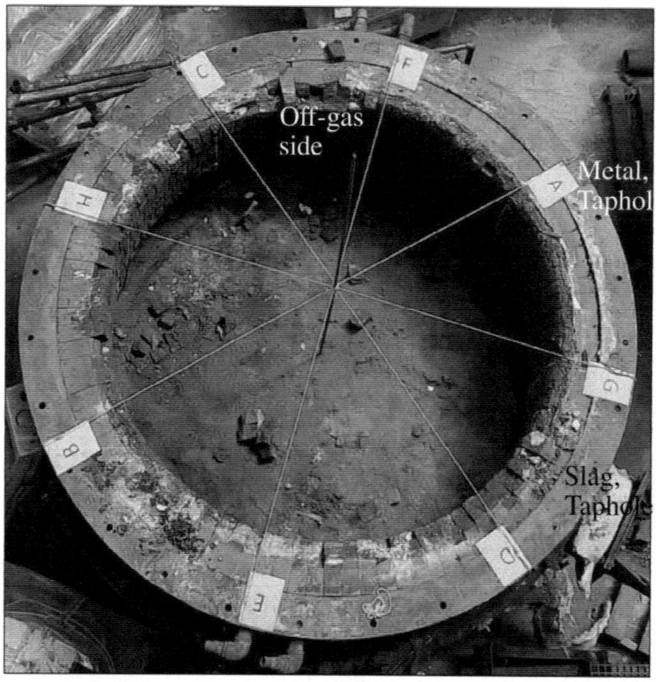


**Figure 14**. Top view of furnace (with roof removed) showing sectioning and measurement of furnace refractory wear post campaign.

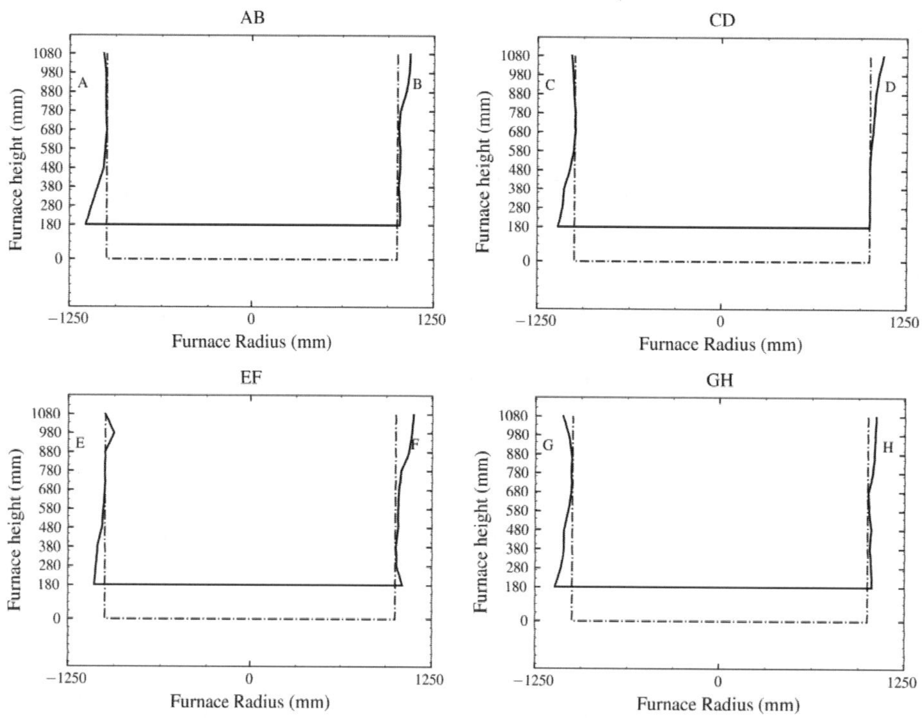


**Figure 15**. Post campaign furnace refractory wear profiles of the different sections.

## Conclusion

Smelting test work was completed in a large pilot-scale 3.2 MVA DC arc furnace to investigate and validate the process efficiencies of the Enviroplas™ process in treating various BOS FC streams to recover Zn and Fe. The campaign processed three different BOS FC streams and a blend of Old and New FC to demonstrate the process by evaluating C and flux requirements, Zn and Fe recovery to respective product streams, product quality, emissions measurements, process energy requirements, and production of environmentally safe slag. The following findings and observations were based on the test work.

- The campaign processed 95 t of BOS FC materials along with an overall addition of 19% (per kg BOS FC) coke, 6.6% silica, and 3.4% limestone. This resulted in an overall metal yield of 38% of the total BOS FC materials processed, and the collection of dust equivalent to 14% of the total mass of materials fed to the furnace. An overall recovery of 89% Zn to the dust and a recovery of 84% Fe to the metal was achieved. However, a deportment of almost 100% Zn to the dust stream was obtained for all BOS FCs smelting conditions while a deportment of Fe to the metal stream was in excess of 90% for all BOS FCs, except for Mixed FC.
- Production of metal with target composition of >96% Fe, <0.5% Si and 2–3% C was successfully demonstrated during smelting of Old, Blend, and Mixed FC. The target metal composition was not demonstrated during smelting of New FC. This condition was characterized by the highest C content (>3%) in metal. This high C content was associated with higher occurrence of other minor impurities such as Cr, Mn, and Si in the metal. This difference in smelting behaviour of New FC was largely due to its phase chemical composition in comparison to the other BOS FC streams.
- As a result of significant variation in slag composition throughout the campaign, the test work was unable to demonstrate production of slag, during smelting of any of the BOS FC, whereby the target composition of 38% $SiO_2$, 25% CaO, 14% MgO, 14% FeO, and 7% $Al_2O_3$, were all met. The Mixed FC condition had the most consistent slag compositions with narrower compositional variation ranges of the components. The weighted average composition was close to target with $SiO_2$ (37.7%), $Al_2O_3$ (5.45%), and CaO (27.7%). Although the composition of MgO was slightly below target, it was the most consistent in all conditions at about 11% while the FeO content had the widest variation of between 3.23 and 18.4%.
- TCLP test results of slag from smelting of each BOS FC stream have shown that the hazardous elements of interest (e.g. Pb, As, and Cd) had leaching rates below 0.1 ppm per unit mass. According to the NSW EPA this classifies all the tested slags as non-hazardous waste. Therefore, the test work demonstrated that the process can produce slags that are environmentally safe for disposal for each BOS FC stream.
- Only Old and Blend FC conditions demonstrated collection of dust with a target composition of >65% ZnO, <20% $Fe_2O_3$, and <2% of PbO, $Al_2O_3$, MgO, MnO, and $SiO_2$. In both conditions, the highest grades achieved had more than 80% ZnO in dust. The New and Mixed FC conditions experienced high feed carryover with the $Fe_2O_3$ averaging 20.6 and 30.3%, respectively. This was largely due to the presence of higher fraction of fines in both streams. The ZnO grade of dust collected during the New and Mixed FC conditions averaged 63.8 and 62.4%, respectively.

Therefore, the project was able to advance the engineering knowledge by further demonstrating and validating the process efficiencies of the Enviroplas™ process in a larger pilot-scale DC arc furnace and increasing the confidence level in the feasibility of the process and implementation at a commercial scale. The test work demonstrated that the Old and Blend steel-dust streams were suitable for processing via the Enviroplas process based on recovered zinc concentration in dust. All feed types are likely to be suitable for processing with minor operational improvements, such as a better feed system to limit carryover of feed dust into the gas stream and efficient control of the bag plant to better define the suction pressure at the gas offtake. It was recommended that the New and Mixed steel-dust streams should be processed as a blend with the Old dust stream to meet the suitability criteria for processing via the Enviroplas process.

## Acknowledgements

This paper is published with the permission of BlueScope. The project was fully funded by BlueScope as part of their development project to find a suitable, cost effective, and environmentally friendly technology to recover valuable Fe and Zn units from various steel-plant dusts.

## References

[1] I. K. Wernick and N. J. Themelis, Recycling metals for the environment, *Annu. Rev. Energy Environ.*, vol. 23, no. 465, p. 97, 1998.
[2] OECD, Government support for primary and secondary metal production, 2018. [Online]. Available: www.oecd.org
[3] E. Matinde, G. Simate, and Ndlovu, Mining and metallurgical wastes: A review of recycling and re-use practices, *J. South. Afr. Inst. Min. Metall.*, vol. **118**, pp. 825–844, Aug. 2018.
[4] R. Nyirenda, The processing of steelmaking flue dust: A review, *Miner. Eng.*, vol. **4**, no. 7, pp. 1003–1025, 1991.
[5] I. Palencia, R. Romero, N. Iglesias, and F. Carranza, Recycling of EAF dust leaching residues to the furnace: A simulation study, *J. Mater.*, pp. 28–32, Aug. 1999.
[6] Z. Bian, X. Miao, S. Lei, S. Chen, W. Wang, and S. Struthers, The challenges of recycling mining and mineral processing wastes, *Science*, vol. **337**, no. 6095, pp. 702–703, 2012.
[7] J. Beukes, P. Van Zyl, and M. Ras, Treatment of Cr (VI)-containing wastes in the South African ferrochrome industry - a review of currently applied methods, *J. South. Afr. Inst. Min. Metall.*, vol. **112**, pp. 347–352, 2012.
[8] X. Lin *et al.*, Pyrometallurgical recycling of electric arc furnace, *J. Clean. Prod.*, vol. 149, pp. 1079–1100, 2017.
[9] K. Mager *et al.*, Recovery of zinc oxide from secondary raw materials: new developments of the Waelz process, in *Proceedings of Recycling of Metals and Engineered Materials*, 2000, pp. 329–344.
[10] Y. Kuwauchi and M. Barati, A mathematical model for carbothermic reduction of dust-carbon composite agglomerates, *ISIJ Int.*, vol. **53**, pp. 1097–1105, 2013.
[11] Y. Hara, N. Ishiwata, H. Itaya, and T. Matsumoto, Smelting reduction process with a coke packed bed for steelmaking dust recycling, *ISIJ Int.*, vol. 50, pp. 1084–1088, 2000.
[12] S. Hughes, M. Reuter, and R. Baxter, Ausmelt technology for lead and zinc processing, *Proc Lead Zinc 2008*, pp. 147–162, 2008.
[13] J. Antrekowitsch, G. Rosler, and S. Steinacker, State of the art in steel mill dust recycling, *Chem Ing Tech*, vol. **87**, pp. 1498–1503, 2015.
[14] M. Kennedy, H. Haaland, and J. Aune, High intensity slag resistance furnace design, in *Pyrometallurgy of Nickel and Cobalt*, 2009.
[15] A. Schoukens, F. Shaw, and E. Chemaly, The Enviroplas process for the treatment of steel-plant dusts, *J. South. Afr. Inst. Min. Metall.*, vol. 93, no. 1, pp. 1–7, 1993.

[16] G. Denton, N. Barcza, P. Scott, and T. Fulton, EAF stainless steel dust processing, in *Sustainable Developments in Metals Processing*, Melbourne, Australia, 2005.
[17] N. Ma, Recycling of basic oxygen furnace steelmaking dust by in-process separation of zinc from the dust, *J. Clean. Prod.*, vol. **112**, no. Part 5, pp. 4497–4504, Jan. 2016, doi: https://doi.org/10.1016/j.jclepro.2015.07.009.
[18] R. Jones and T. Curr, Pyrometallurgy at Mintek, in *Southern African Pyrometallurgy*, 2006, pp. 127–150.
[19] EPA, Method 1311 - Toxicity characteristic leaching procedure. Environmental Protection Agency, Jul. 1992. Accessed: May 20, 2020. [Online]. Available: https://www.epa.gov/sites/production/files/2015-12/documents/1311.pdf
[20] NSW Environment Protection Authority, Waste Classification Guidelines - Part 1: Classifying waste, NSW EPA, Sydney, Nov. 2014. [Online]. Available: www.epa.nsw.gov.au
[21] DFFE, *National Environment Management: Air Quality Act (no. 39 of 2004)*, vol. 476. 2005.
[22] NSW Environment Protection Authority, *Protection of the Environment Operations Act 1997*, vol. 811. 2022.

Journal of Physics: Conference Series **3050** (2025) 012006 doi:10.1088/1742-6596/3050/1/012006

# Mass and energy based modelling of EAF steelmaking scenarios using scrap and hydrogen reduced DRI as raw materials

**Eetu-Pekka Heikkinen[1*], Petri Sulasalmi[1], Ville-Valtteri Visuri[1], Seppo Ollila[2] and Jarmo Lilja[2]**

[1] Process Metallurgy Research Unit, University of Oulu, Oulu, Finland
[2] SSAB Europe Oy, Raahe, Finland

[*]E-mail: eetu.heikkinen@oulu.fi

**Abstract.** Pressure for radical reduction of $CO_2$ emission in steel industry favours EAF steelmaking using scrap and hydrogen reduced iron (H2DRI) as raw material over traditional blast furnace-based production. The increase in EAF-based production will undoubtedly affect on demand of raw materials. However, the availability of H2DRI will not allow 100% H2DRI-based production immediately. On the other hand, increasing share of EAF production in steelmaking yields increase in scrap demand. Therefore, different scenarios of steelmaking are likely to appear where variable mixture of scrap and H2DRI will be used in EAF. The ratio of different raw materials (scrap and H2DRI) will affect various process parameters such as the quality of molten steel, energy consumption of the process, amount of slag, $CO_2$ emissions. Mass and energy balance modelling provides a suitable way of studying possible future EAF steelmaking concepts. In this study, a set of scenarios were studied using a mass and energy balance based model with HSC Sim thermodynamic software. The model as well as the studied scenarios were constructed based on data available in the literature supported with industrial and experimental data. The main interest in the study was on energy consumption and $CO_2$ emissions when using scrap and H2DRI in multiple ratios and in different temperatures as a raw material.

## 1. Introduction

Steel industry's ambitions towards fossil-free production require significant changes in the existing processes as conventional blast furnace (BF) – basic oxygen furnace (BOF) -route is being replaced by the production of sponge iron (DRI) combined with smelting in either electric arc furnace (EAF) or other electric furnace. Due to potential limitations on the availability of DRI it is likely to use different kind of mixtures of DRI and steel scrap as raw materials for the EAF. These variations in the raw materials have an effect on the composition and quality of produced steel, amount and composition of slag as well as energy consumption and emissions.

Numerical models have been developed to simulate various aspects of the EAF process [1, 2]. Mass and heat balance models are ideal tools for conducting scenario analyses in cases where experimental data is not yet available. Due to their relatively simple kinetic descriptions and lower computational expense, these models are better suited for simulating larger system boundaries than typical dynamic process models.

The aim of this study was to create a mass and energy balance based model for EAF and its support processes and use this model to simulate the effect of raw materials on the products, emissions and energy consumption of the furnace.

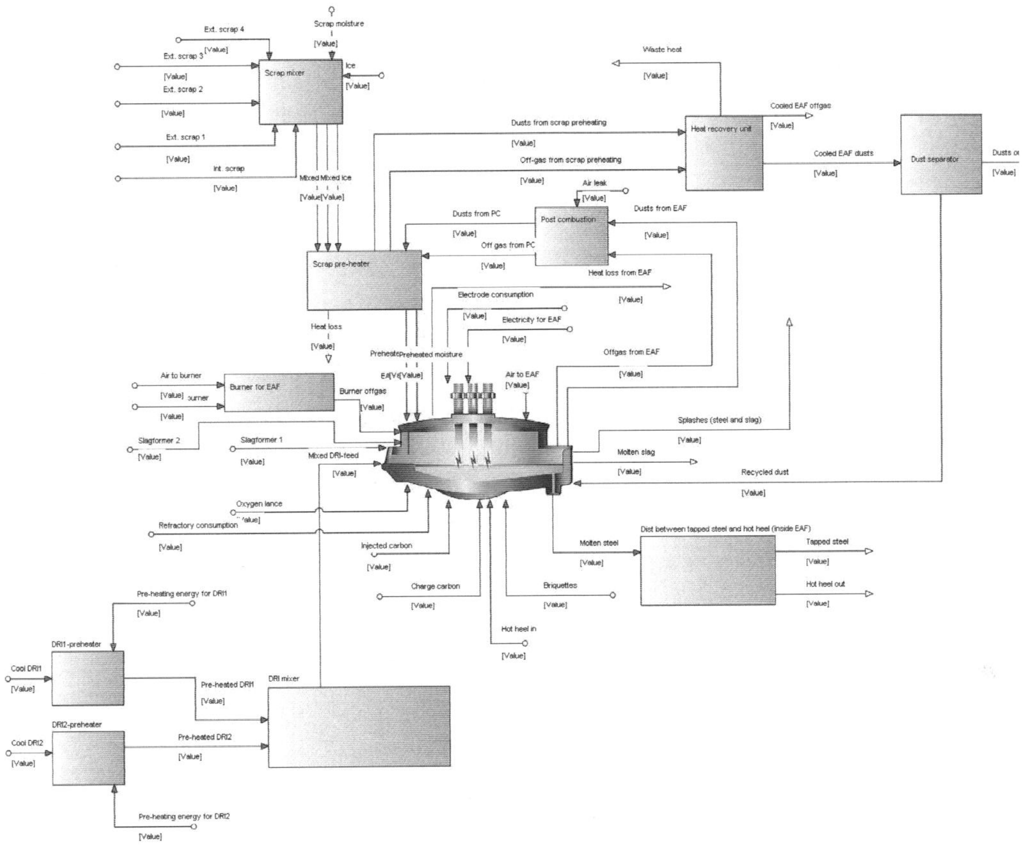


**Figure 1.** Flowsheet of the used process model created with HSC Sim.

## 2. Modelling and simulations

A mass and energy balance based process model for the electric arc furnace and its support processes was created with HSC Sim -software. Support processes included in the model were preheaters and mixers for both DRI and steel scrap as well as offgas treatment units (*i.e.*, post combustion unit, heat recovery unit and dust separator). Additionally, a burner and a distributor of steel melt between tapped steel and a hot heel were considered as separate process units in the model. The flowsheet of the process model is shown in Figure 1. All the process units shown in Figure 1 were modeled using a Pyro unit -option of the software.

List of parameters needed for the simulations are summarised in Table 1. The values for these parameters could be changed, but they (apart from the variables mentioned in Table 2) were held constant in the simulations presented in this paper. The values for the simulations were obtained from the literature as well as from the measurements and analysis results obtained from the industrial partners. List of 14 scenarios studied in this paper together with used variables is doi:10.1088/1742-6596/3050/1/012006

presented in Table 2. For all these scenarios, only one ($H_2$-reduced) DRI grade with a metallization degree of 97% and a constant composition was considered. Chemical compositions of steel scrap mix, recycling briquettes, hot heel as well as injected carbon were also held constant. The amounts of oxygen injection and injected carbon were considered to be 0.9 $Nm^3$/t-Fe and 15.0 kg/t-Fe, respectively. No charged carbon or doloma as a slag forming agents were used in these scenarios. Slag basicity and FeO-content were set as 2.5 and 25.0 wt-%, respectively, whereas the C-content in steel was set as 0.04 wt-%. All the simulations were made for 220 ton heats with a 60 wt-% hot heel left for the following heat. As seen from Table 2, some variables, which could have been varied in the model, were held constant in the scenarios presented in this paper. For example, temperature of scrap was set to 20°C in all the presented cases, although it is possible to take scrap preheating into account in the simulations.

**Table 1.** Parameters needed as initial values for the simulations.

| Raw materials | Products | Process parameters & Others |
|---|---|---|
| Amounts and chemical compositions of five different steel scraps | Target C and Si in steel | Burner: Amounts and chemical compositions of fuel and air/oxygen. Air/Fuel -ratio (lambda). |
| Amounts and chemical compositions of two different DRI pellets | Amount of tapped steel | Refractory materials: Chemical composition. Consumption. |
| (DRI/Scrap -ratio) | Target basicity ($CaO/SiO_2$) and FeO-content in slag | Electrodes: Chemical composition. Consumption. |
| Amounts and chemical compositions of recycled briquettes | Temperatures of steel, slag, splashes, dust and offgas (before and after cooling). | Amount of splashes (steel, slag) and mechanically formed dust. |
| Amounts and chemical compositions of injected and charge carbon | | Hot heel: Amounts and chemical compositions (from the previous heat / to the next heat). Temperature loss between the heats. |
| Compositions of two different slag forming agents (lime, doloma) | | Air leakage: Amounts and chemical compositions (EAF and post combustion). |
| Temperatures of all raw materials (without preheating) | | Heat losses of EAF and support processes. |
| Temperatures of DRI and steel scrap (after preheating) | | |
| Oxygen lancing. | | |

**Table 2.** Scenarios studied in this paper.

| Scenario | Scrap | | Hot DRI (HDRI) | | Cold DRI (CDRI) | |
|---|---|---|---|---|---|---|
| | [wt-%] | [°C] | [wt-%] | [°C] | [wt-%] | [°C] |
| 1 | 100.0 | 20 | 0.0 | - | 0.0 | - |
| 2 | 89.6 | 20 | 0.0 | - | 10.4 | 20 |
| 3 | 65.4 | 20 | 34.6 | 200 | 0.0 | - |
| 4 | 65.4 | 20 | 34.6 | 400 | 0.0 | - |
| 5 | 65.4 | 20 | 34.6 | 600 | 0.0 | - |
| 6 | 55.0 | 20 | 34.6 | 200 | 10.4 | 20 |
| 7 | 55.0 | 20 | 34.6 | 400 | 10.4 | 20 |
| 8 | 55.0 | 20 | 34.6 | 600 | 10.4 | 20 |
| 9 | 55.0 | 20 | 45.0 | 200 | 0.0 | - |
| 10 | 55.0 | 20 | 45.0 | 400 | 0.0 | - |
| 11 | 55.0 | 20 | 45.0 | 600 | 0.0 | - |
| 12 | 44.6 | 20 | 45.0 | 200 | 10.4 | 20 |
| 13 | 44.6 | 20 | 45.0 | 400 | 10.4 | 20 |
| 14 | 44.6 | 20 | 45.0 | 600 | 10.4 | 20 |

Concerning the behavior of elements in the EAF, the following assumptions were made:

- Aluminium, calcium, magnesium and titanium end up completely to the slag (apart from portions in mechanically formed dust and splashes).
- Copper, molybdenum, nickel and lead end up completely to the steel (apart from portions in mechanically formed dust and splashes).
- Hydrogen and nitrogen end up completely to the gas phase.
- Chromium, manganese, phosphorus, sulphur and vanadium are distributed between steel and slag according to the defined ratios. For chromium and vanadium, this ratio was assumed to be constant, for sulphur the ratio depended on the slag basicity, for manganese the ratio depended on the slag's FeO-content and for phosphorus the ratio depended on the slag basicity and FeO-content as well as temperature.
- A constant portions of potassium, sodium and zinc end up to the slag and the rest ends up in the dust via gas phase (*i.e.*, vaporisation).
- Carbon content in steel is set to a constant value and the rest end up in the gas phase (apart from portions in mechanically formed dust and splashes).

- Silicon content in steel is set to a constant value and the rest end up in the slag (apart from portions in mechanically formed dust and splashes).
- Iron content in slag (as FeO) is set to a constant value and the rest end up to the steel (apart from portions in mechanically formed dust and splashes).
- Oxygen is used to fix the material balances (by "Float"-option in HSC Sim).

As a result of simulations, the following data was obtained for each scenario:

- Amounts and chemical compositions of output streams; *i.e.,* tapped steel, slag, offgas (before and after cooling) dust (before and after cooling), splashes and hot heel to the following heat.
- Amounts of input streams (for which the compositions were given as initial values); *i.e.,* DRIs, steel scraps, recycling briquettes, slag forming agents (lime, doloma), injected and charged carbon, oxygen feed via lance, hot heel from the previous heat as well as used fuel and air or oxygen in the burner.
- Amounts of consumed refractory materials and electrodes.
- Amounts of air leakage to the EAF and post combustion.
- Need for electricity.
- Heat produced by the burners.
- Heat used in the preheating of DRI and/or steel scrap.
- Heat losses from EAF, preheating and gas treatment.
- Total direct $CO_2$ emissions.

The results could be presented either as tons (or kgs) per heat or tons (or kgs) per produced ton of steel.

## 3. Results and discussion

### *3.1 Output material streams*

The amounts of the following output material streams were obtained for all 14 scenarios: tapped steel, hot heel to the following heat, slag, splashes as well as cooled offgas and dusts. All scenarios were calculated for a charge weight of 220 tons. The amounts of hot heel and splashes were assumed not to change between the scenarios and their amounts were 132 t/heat (600 kg/t-Fe) and 2.2 t/heat (9.9 kg/t-Fe), respectively.

Simulated amounts of slag (kg/t-Fe), cooled dust (kg/t-Fe) and cooled offgas ($Nm^3$/t-Fe) for all 14 scenarios are presented in Figures 2, 3 and 4, respectively. It is seen that the amount of slag is increased with increasing DRI/scrap-ratio due to DRI's larger amount of harm minerals and increased need of slag forming agents in order to keep the slag basicity on a wanted level. For scenarios 6-14 the amount of slag is more than two times larger in comparison to 100% scrap charge. According to the simulations the amounts of slag varied between 21.9 kg/t-Fe (in scenario 1) and 47.2 kg/t-Fe (in scenarios 6-8). These values are considerably lower in comparison to the typical slag quantities of 60-270 kg/t-Fe in European electric arc furnaces [3, p. 429]. In the model, the amount of slag is mainly defined by the amount of lime and/or doloma required to keep the slag basicity on a defined level (which is 2.5 in the scenarios of this study). Further on, the amounts of these slag forming agents are defined based on how much the raw materials contain elements (mainly Si) that eventually end up in the slag phase either by oxidation during the process or by occuring as oxides already in the input materials. Since the purity of the raw materials was considered to be relatively high in the scenarios of this study, the required amounts of slag forming agents were also smaller (11-14 kg/t-Fe) in comparison to the typical values of the industry (25-140 kg/t-Fe [3, p. 429]), which in turn leads into smaller amounts of slag.

The amount of dusts also increased with increasing DRI/scrap-ratio, although the effect of raw material ratio is significantly smaller in comparion to the slag. The amounf of dusts varied between 42.3 kg/t-Fe (in scenario 1) and 43.1 kg/t-Fe (in scenarios 6-14). When compared with the typical values of dust formation in European electric arc furnaces, these values are slightly higher than the ones presented in e.g. BAT reference documents (10-30 kg/t-Fe [3, p. 429]). The amount of dust formation in the model is solely defined by a so-called "dusting coefficient", which defines the portion of each element ending up in the dust. This coefficient was assumed to be 2.5% for all elements apart from sulfur, nitrogen, sodium, potassium and zinc, which are assumed to end up in the gas (with small portions of Na, K and Zn also to the slag), as well as oxygen, which is used as a so-called floating element. This value was used, since it had proved to give good correspondance with industrial values in the previous models and simulations of comparable systems. It can be argued that a slightly smaller value could have given more realistic values in the scenarios of this study.

Unlike the amounts of slag and dusts, the amount of offgas decreased with increasing DRI/scrap-ratio varying between 357 $Nm^3$/t-Fe (in scenarios 9-11) and 379 $Nm^3$/t-Fe (in scenario 1). The decreasing effect can be explained with the lower C-content in the DRI-pellets in comparison to the steel scrap. Due to this less $CO_2$ is formed and the amount of offgas is decreased. This is also seen in the simulated $CO_2$ emissions presented in Figure 5. The specific $CO_2$ emissions are decreased from 67.4 kg/t-Fe (in scenario 1) to 64.2 kg/t-Fe (in scenarios 6-14), when the share of DRI is increased in the charge. The simulated $CO_2$ emissions are slightly lower than the typical values of European electric arc furnaces (72-180 kg/t-Fe [3, p. 429]). Due to the assumptions concerning the behavior of carbon (*i.e.* constant concentration in steel and rest oxidized into the gas phase), the amount of $CO_2$ emissions depends solely on the amounts of carbon in input materials (*i.e.* scrap, DRI, recycled briquettes, injected and charged carbon as well as electrode wear) and hot heel remaining from the previous heat. The lower these amounts are, the lower are the $CO_2$ emissions, too.

Generally speaking, one should keep in mind the assumptions made for behavior of different elements when concerning the applicability of the simulated results into the practice. When comparing the distribution of elements between metal and slag (or metal and gas), it is reasonable to assume that some elements with very high affinity for oxygen (such as aluminium, calcium and magnesium) end up solely in the slag, whereas some elements with low affinity for oxygen (such as copper) end up solely in the metal. In addition to these, there are some elements with relatively low boiling point (such as zinc and the alkali), which end up solely in the gas (and eventually in the dusts after gas has cooled down). However, for many elements it is necessary to define some kind of distribution between the phases. This can be done by assuming a constant content in one of the phases (in which case the rest ends up in the other phase) or by assuming a ratio for contents in two different phases. This ratio can be defined either as a constant or it could be a function of varying conditions (such as temperature, partial pressure of oxygen, slag basicity, etc.).

From the results' applicability's perspective, the assumption of constant contents is justified as long as the steel and the slag that are being produced have more or less constant composition (*i.e.* there are no big variations on the steel grades whose production is being simulated). In our simulations, this kind of assumption was made for carbon (between metal and gas) and silicon (between metal and slag), because the produced steel grade was not considered as a variable in the scenarios of this study. However, if the model is used to simulate the production of steels and/or slags or greater variety, this kind of assumption is not valid and it is more appropriate to use distribution ratios based on thermodynamic equilibria or other data describing the relevant reactions between the phases. In the current study this kind of assumptions were made for chromium, manganese, phosphorus, sulphur and vanadium.

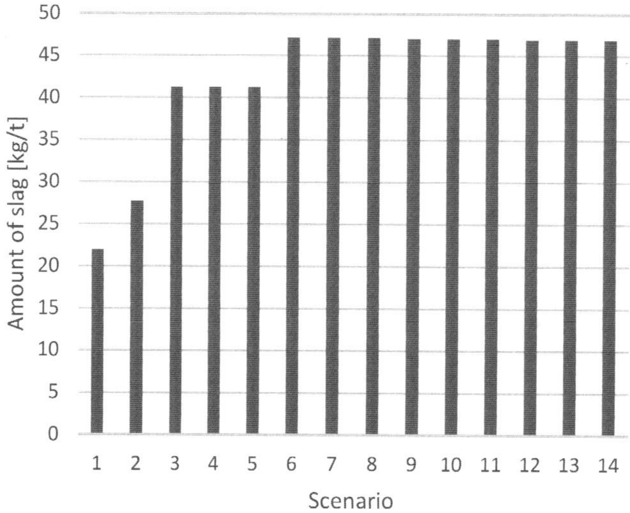


**Figure 2.** Simulated amount of slag (kg/t-Fe) for 14 studied scenarios.

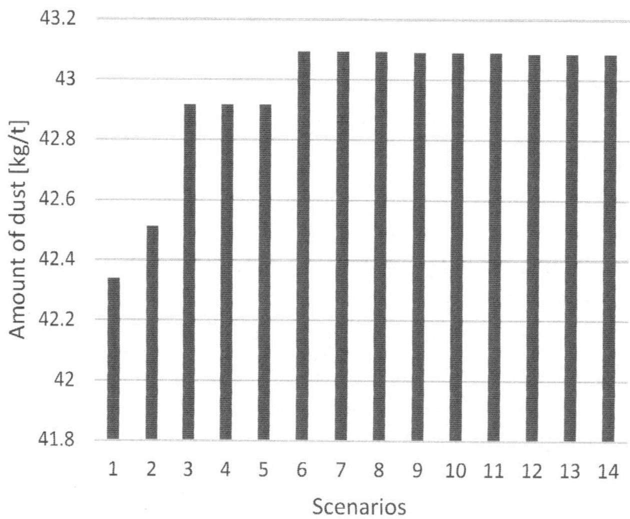


**Figure 3.** Simulated amount of cooled dust (kg/t-Fe) for 14 studied scenarios.

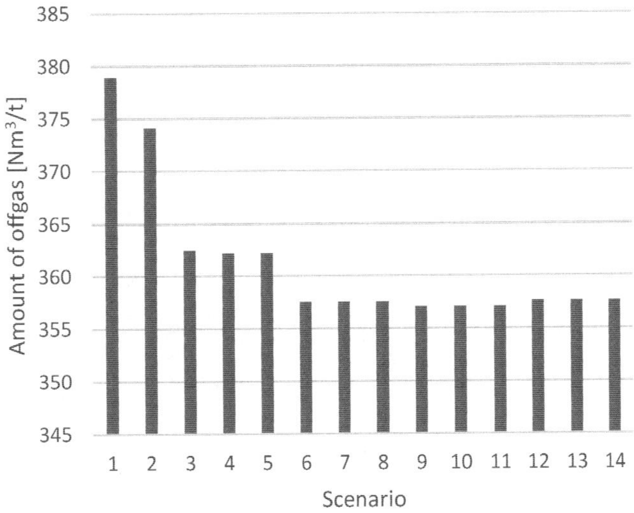


**Figure 4.** Simulated amount of cooled offgas ($Nm^3$/t-Fe) for 14 studied scenarios.

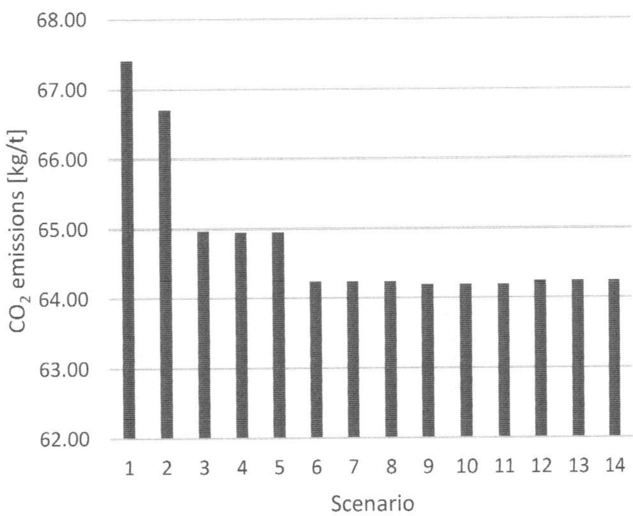


**Figure 5.** Simulated direct $CO_2$ emissions (kg/t-Fe) for 14 studied scenarios.

*3.2 Energy consumption*

Simulated total energy need (kWh/t-Fe) for all 14 scenarios is presented in Figure 6. It is seen that required total energy is increased as the DRI/scrap-ratio is increased. This is mainly due to increased amount of slag (*cf.* Figure 2): more energy is needed to heat and smelt larger amounts of slag. As seen from Figure 6, the total energy need varied between 467 kWh/t-Fe (in scenario 1) and 499 kWh/t-Fe (in scenarios 6-14).

Simulated electricity requirements (kWh/t-Fe) and energies needed in the preheating of the DRI charge (kWh/t-Fe) are presented in Figures 7 and 8, respectively. As more energy is used for DRI preheating (Figure 8), less electricity is needed (Figure 7). As expected, the lowest electricity consumption is seen in scenario 11, which has the largest share of HDRI and highest temperature of preheated DRI.

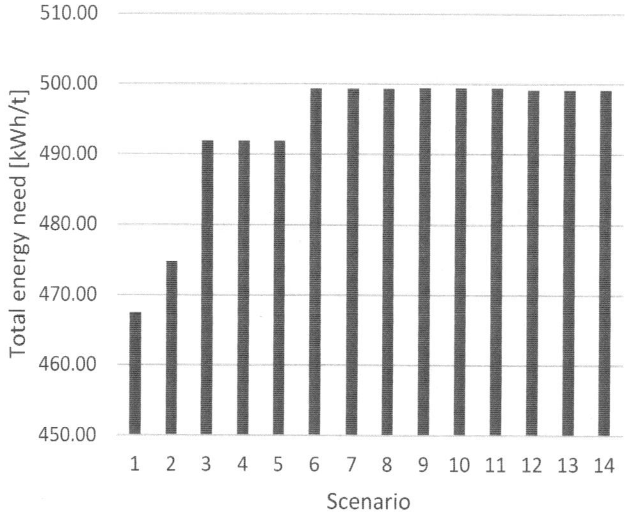


**Figure 6.** Simulated total energy need (kWh/t-Fe) for 14 studied scenarios. Total energy includes EAF electricity, burners as well as DRI preheating.

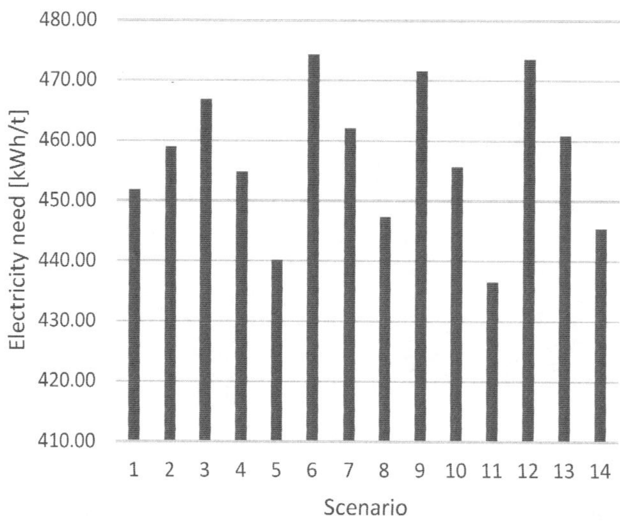


**Figure 7.** Simulated electricity need (kWh/t-Fe) in the EAF for 14 studied scenarios.

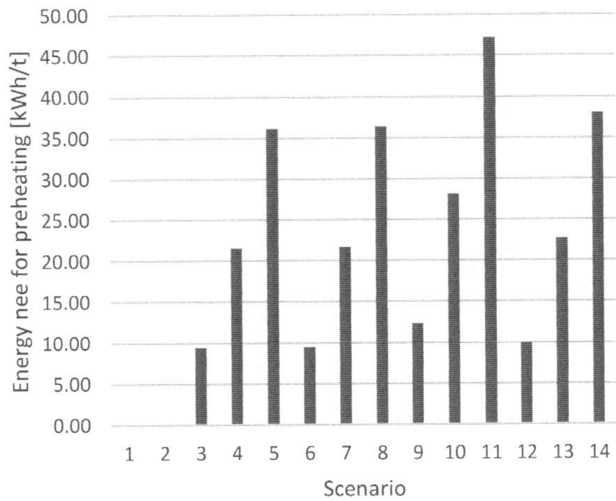


**Figure 8.** Simulated energy need for DRI preheating (kWh/t-Fe) for 14 studied scenarios.

*3.3 Effect on the steel composition*

Other scenarios were also tested with the EAF model. For example, after the addition of more detailed description of certain elemental distributions, the model has been used to estimate the behavior of tramp elements and the effect of different raw materials on the composition of produced steel. The results of these simulations enable the estimation of which DRI/scrap-ratios can be used to keep the impurity levels under the levels required for produced steel grades. [4]

## 4. Summary

Mass and energy balance based model was created with HSC Sim -software and the model was used to simulate scenarios, in which different mixtures of raw materials (steel scrap, HDRI, CDRI) were used. Despite some rather simplistic assumptions on the behavior of certain elements, it is possible the estimate the effect of different raw materials on the products, by-products and emissions as well as on the energy consumption of the process based on the simulation results. The structure of the model enables the re-definition of the behavior of elements thus enabling the simulation of steel composition more accurately.

## Acknowledgements

Development of the model was conducted within the FFS2 project (5534/31/2023) funded by Business Finland.

## References

[1] Hay T, Visuri V-V, Aula M & Echterhof T. 2021. A Review of Mathematical Process Models for the Electric Arc Furnace Process. *Steel Res. Int.* **92**. 2000395.

[2] Odenthal H-J, Kemminger A, Krause F, Sankowski L, Uebber N & Vogl N. 2018. Review on Modeling and Simulation of the Electric Arc Furnace (EAF). *Steel Res. Int.* **89**. 1700098.

[3] Remus R, Aguardo-Monsonet MA, Roudier S & Delgado Sancho L. 2013. Best Available Techniques (BAT) Reference Document for Iron and Steel Production. *Industrial Emissions Directive 2010/75/EU. JRC Reference Report.* European Commission.

[4] Mehtälä M. 2023. Management of metallic impurities in production of strip steels in EAF process. *MSc Thesis.* University of Oulu, Finland.

# Kinetic Modeling of Hematite Reduction by Hydrogen Plasma Smelting Reduction in Laboratory–Scale

**A Javed[1*], I Mäkelä[1], H Pauna[1], H-R Putaala[1], U Manzoor[2], D Klapproth[2], I R Souza Filho[2,3], V-V Visuri[1]**

[1]Process Metallurgy Research Unit, University of Oulu, P.O. Box 4300, 90014, Oulu, Finland.
[2]Max Planck Institute for Sustainable Materials, Max-Planck-Str. 1, 40237, Düsseldorf, Germany.
[3]Institut Jean Lamour, Université de Lorraine, 2 allée André Guinier, 54011, Nancy, France.

*Email: areej.javed@oulu.fi

**Abstract.** The steel industry accounts for about 8% of global $CO_2$ emissions, primarily due to carbon–based compounds like coke for iron ore reduction, highlighting the importance of developing fossil–free reduction processes. Hydrogen–based reduction has emerged as a promising alternative to reduce $CO_2$ emissions. In addition to hydrogen direct reduction, hydrogen plasma smelting reduction (HPSR) has been envisaged as a one–step process for melting, reducing, and refining steel from iron ore. Mathematical models and new monitoring methods are needed to control and predict the process. To this end, this work aimed to develop a kinetic model for the reduction of hematite ($Fe_2O_3$) to explore the dynamic interactions and reaction mechanisms in the HPSR process. The gas–metal–slag model is used to calculate the reactions occurring between the molten metal, liquid slag, and gas phase, using the effective equilibrium constant (EEC) approach, focusing on the stepwise reduction of hematite via magnetite ($Fe_3O_4$) and wüstite (FeO) to metallic iron (Fe). The model can predict (1) the dynamic changes in the composition of the metal and oxide phases during the reduction process (2) the relative reduction efficiencies of intermediate iron oxides under specific plasma conditions. The model predictions are validated using data from laboratory–scale experiments. The model predictions for the kinetic behavior of the system were found to be in good agreement with the experimental results.

## 1. Introduction

Iron ore reduction using hydrogen plasma is a promising technique to produce sustainable steel and to reduce the $CO_2$ emissions associated with the traditional iron and steel producing processes. On average, the steel industry contributes around 7–9% of global $CO_2$ emissions, mostly due to the use of coke as a reducing agent [1]. To avoid these emissions, hydrogen-based reduction technologies have gained considerable attention to produce green steel [2].

Direct injection into blast furnaces [3], direct reduction in shaft furnaces or fluidized bed reactors [4], and the HPSR process [5-7] are the three primary methods for producing iron utilizing hydrogen as a reducing agent that are taken into consideration. The first method is technologically advanced but limiting as it requires coke as support. Direct reduction of iron ore using hydrogen can be carried out, e.g., in a shaft furnace or fluidized bed reactor, producing solid sponge iron that must then be melted in an electric arc furnace. Hydrogen plasma–based reduction is thermodynamically more efficient and allows for simultaneous reduction and melting, but it is still at a relatively low technology readiness level. In addition to molecular hydrogen, atomic and ionic hydrogen species are present in the plasma, allowing a more efficient reduction and melting of iron ores [7]. As the reactions are exothermic for both monoatomic (H) and ionized hydrogen ($H^+$) and produce a high amount of heat at the surface of the reaction, the need for external power is reduced, and the overall efficiency of the process increases [6].

Thermodynamic calculation conducted by Seftejani et al. [5], suggests that high temperatures in the plasma arc ionize the $H_2$ into atomic and ionized hydrogen, and atomic hydrogen is the most effective agent for the reduction of ores. Nagasaka et al.[8], presented that the possible rate–limiting steps for the reduction of liquid iron oxide with molecular hydrogen are mass transfer in the gas phase, interfacial reaction, and mass transfer of oxygen in the liquid phase to the interface, whereas Zarl et al. [9], stated that at high temperatures and in the presence of multiple phases, i.e., liquid phase and plasma, mass transport is the rate–limiting step. Souza Filho et al.[7], highlighted that optimizing input materials significantly enhanced kinetics, with hematite fully transforming to wüstite within one minute of hydrogen plasma exposure. However, an additional 14 minutes were required for complete reduction to metallic iron, with wüstite reduction being the rate-limiting step. Zarl et al. [9], proposed a four–phase kinetic model of hydrogen plasma reduction, categorizing the process into premelting, pre–reduction, reduction and charging, and post–reduction stages, while recommending further studies to explore process influences, especially during the charging phase. Ernst et al. [10], examined the reduction behavior of various iron ores. They found that magnetite exhibited the highest oxygen removal rate, while hematite had the steepest reduction progression due to its high oxygen content. Additionally, a drop-in reduction rate above 80–85% reduction degree suggested a change in kinetics. Comparison of the hematite pellet with the powder form shows that the powder exhibits ten times faster reduction kinetics than in the pellet form [11].

These studies collectively contribute to understanding HPSR by exploring key aspects such as reduction kinetics, phase transformations, arc stability, and hydrogen utilization efficiency based on empirical observations and experimental findings. However, despite these studies, a dynamic model that can predict the kinetics of iron ore reduction and the system's mass and energy balance is still lacking, which is required for optimizing and scaling up the HPSR process.

In this paper, we aim to present a kinetic model to analyze the intricate interactions and reaction mechanisms occurring during the reduction of hematite via hydrogen plasma on a lab scale. This model adopts a gas–metal–slag framework based on an effective equilibrium constant (EEC) approach to simulate the reduction of hematite to magnetite, wüstite, and ultimately, metallic iron.

## 2. Model description

The developed gas–metal–slag model is a modified version of the metal–slag module developed initially by Mäkelä et al. [12], based on the effective equilibrium constant method proposed by Richardson et al. [13] and its modification by Ohguchi et al. [14]. In the model, the term slag refers to the unreacted iron oxides present in the system. The model describes the reactions occurring between the molten metal and liquid slag in the presence of the hydrogen plasma. Metal–slag reactions are complex equilibrium processes involving multiple components, where various elements interact simultaneously. Competitive oxidation and reduction reactions form a mass transfer–controlled equilibrium at the reaction interface. The equilibrium at the interface is solved based on effective equilibrium constants, for which equilibrium constants, activities of species in the metal phase, activities of species in the oxide phase, and partial pressures of gases are needed. Regular solution model (RSM) by Ban-Ya [15] and Unified Interaction Parameter Formalism (UIPF) [16] are employed to estimate the activities of species in the molten oxide and metal phases. Because the system operates at close-to-atmospheric pressure, the species in the gas phase are treated according to ideal law. Hence, their activities are substituted with their partial pressures.

The original model by Mäkelä et al. [12] considered three reaction types: oxidation reactions "ox", ion exchange reactions "ex", and reactions for the carbon monoxide production reaction as "gas". In this work, a new reaction type is considered that defines the reactions between the plasma phase ($H_2$, H) and the liquid melt phase [O], named as "2gas". The water vapor produced is considered to leave the system without affecting other reactions. The chemical reactions considered in this study are shown in equations (1)–(5), where [], (), and {} denote species in the metal, oxide, and gas phases. $FeO_{1.5}$ and $FeO_{1.33}$ notations are used for $Fe_2O_3$ and $Fe_3O_4$, respectively, in the model description.

$$\mathrm{Fe} + [\mathrm{O}] = (\mathrm{FeO}_{1.5}), \tag{1}$$
$$\mathrm{Fe} + [\mathrm{O}] = (\mathrm{FeO}_{1.33}), \tag{2}$$
$$\mathrm{Fe} + [\mathrm{O}] = (\mathrm{FeO}), \tag{3}$$
$$\{\mathrm{H}_2\} + [\mathrm{O}] = \{\mathrm{H}_2\mathrm{O}\}, \tag{4}$$
$$\{\mathrm{H}\} + 2[\mathrm{O}] = 2\{\mathrm{H}_2\mathrm{O}\}. \tag{5}$$

The equilibrium constant of each reaction is calculated using the following equation:

$$K = \exp\left(-\frac{\Delta G^{\mathrm{o}}}{RT}\right), \tag{6}$$

$K$ represents the equilibrium constant, $\Delta G^{\mathrm{o}}$ is the change in the standard Gibbs free energy of the reaction, incorporating dissolution effects where relevant [$\mathrm{J \cdot mol^{-1}}$], $R$ is the universal gas constant [$\mathrm{J \cdot mol^{-1} \cdot K^{-1}}$], and $T$ is the interfacial temperature in Kelvin. The model derives $K$ based on reactions defined in the system. A key distinction in this work is the explicit addition of a separate plasma phase, where plasma–phase species react with dissolved oxygen [O] in the melt at the plasma–melt interface. In this approach, for the plasma reactions, the equilibrium constants are computed by considering only the Gibbs energy of dissolution for [O], while the Gibbs energy of dissolution of the plasma species are not included. The calculated equilibrium constants were transformed into the effective equilibrium constants for iron oxides, based on the work of Mäkelä et al. [12] and Ohguchi et al. [14] as shown in equation (7).

$$E_{\mathrm{i}} = \frac{(\%iO_{\mathrm{x}})^*}{a_{[\mathrm{O}]^*}{}^{\nu_{\mathrm{O}}}} = \frac{100\, C\, M_{\mathrm{iO_x}}\, a_{\mathrm{Fe}}\, K_{\mathrm{i}}}{\rho_{\mathrm{iO_x}}\, \gamma_{\mathrm{iO_x}}}, \tag{7}$$

where $E_{\mathrm{i}}$ is the effective equilibrium constant for the reaction forming species $i$, $\nu_{\mathrm{O}}$ is the stoichiometric coefficient of oxygen in oxide $i$, $a_{[\mathrm{O}]^*}$ is the activity of the oxygen, $a_{\mathrm{Fe}}$ is the activity of iron, $\gamma_{\mathrm{iO_x}}$ is the activity coefficient of the species $i$, $K_{\mathrm{i}}$ is the equilibrium constant for the reaction, $M_{\mathrm{iO_x}}$ is the molecular weight of the oxide species, $(\%iO_{\mathrm{x}})^*$ is the weight percentage of species $i$ at the interface, $\rho_{\mathrm{iO_x}}$ is the density of the oxide species, and $C$ is the total molar concentration of the slag [$\mathrm{mol \cdot m^{-3}}$].

Journal of Physics: Conference Series **3050** (2025) 012007 doi:10.1088/1742-6596/3050/1/012007

After determining the effective equilibrium constant, the interfacial weight percentages of the slag components were calculated. To calculate the interfacial partial pressure of the gas phase, the partial pressure of the bulk gas phase is first determined using the ideal gas law,

$$P_\mathrm{i} = y_\mathrm{i}\, P, \tag{8}$$

where, $P_\mathrm{i}$, is the partial pressure of each gas species present in the gas phase [atm], $y_\mathrm{i}$ is the molar fraction of each species, and P is the total pressure of the system [atm]. Once the bulk partial pressures are calculated the interfacial partial pressures are calculated using the following equation:

$$P_\mathrm{i}^* = \frac{P_\mathrm{i}^\mathrm{b} + \left(P_{\mathrm{H_2O}}^\mathrm{b}\, \frac{G_{\mathrm{H_2O}}}{G_\mathrm{i}}\right)}{\left(1 + \frac{G_{\mathrm{H_2O}}}{G_\mathrm{i}}\, E_\mathrm{i}\, a_{[\mathrm{O}]^*}\right)}, \tag{9}$$

where $P_\mathrm{i}^*$ is the interfacial partial pressure of each gas species that is taking part in "2gas" reaction, $P_\mathrm{i}^\mathrm{b}$ is the bulk partial pressure of each gas, $G_{\mathrm{H_2O}}$ and $G_\mathrm{i}$ are the rate constants for the "2gas" reactions production and consumption [mol·m$^{-2}$·sec$^{-1}$]. $E_\mathrm{i}$ is the effective equilibrium constant for the respective reaction, and $a_{[\mathrm{O}]^*}$ is the activity of the dissolved oxygen at the interface. The oxygen activity at the interface can be solved based on the electroneutrality condition. The oxygen activity equation considers the flux density of the metal species along with the flux density of the slag species to consider the contribution of different oxides present in the system. The $a_{[O]^*}$ is solved numerically by using the Newton's method for which differentiates were calculated numerically using the first order forward differentiation method.

$$f(a_{[\mathrm{O}]^*}) = \sum \nu_{\mathrm{O,j}}\, J_{\mathrm{j,ox}} - \sum J_{\mathrm{j,ex}} + \sum \nu_{\mathrm{O,j}}\, J_{\mathrm{j,gas}} + \sum J_{\mathrm{j,2gas}} - F_\mathrm{O}\{[\%\mathrm{O}]^\mathrm{b} - a_{[\mathrm{O}]^*}\} = 0, \tag{10}$$

where, $\nu_{\mathrm{O,j}}$ is the stoichiometric coefficient of oxygen in reaction $j$, $J_{\mathrm{j,ox}}$ is the flux density from oxidation reactions [mol·m$^{-2}$·s$^{-1}$], $J_{\mathrm{j,ex}}$ is the flux density from ion exchange reactions [mol· m$^{-2}$·s$^{-1}$], $J_{\mathrm{j,gas}}$ is the flux from reactions that generate gases [mol·m$^{-2}$·s$^{-1}$], $J_{\mathrm{j,2gas}}$ is the flux density from gas–phase reactions [mol·m$^{-2}$·s$^{-1}$], $[\%\mathrm{O}]^\mathrm{b}$ is the bulk concentration of oxygen in liquid melt, and $a_{[O]^*}$ is the interfacial activity or concentration of oxygen at the interface. Using the interfacial weight percentages and $a_{[O]^*}$, modified mass transfer of each component in the metal and slag phase, and concentration change with small time step Δt was calculated. For the gas phase reactions, the concentration change was calculated based on the arbitrarily defined rate constant for each "2gas" reaction. The flux density of each phase is calculated using the following equations:

$$J_\mathrm{i} = \frac{\beta_\mathrm{m}\, \rho_\mathrm{m}}{100\, M_\mathrm{i}} ([\%i]^\mathrm{b} - [\%i]^*), \tag{11}$$

$$J_\mathrm{i} = \frac{\beta_\mathrm{s}\, \rho_\mathrm{s}}{100\, M_{\mathrm{iO_x}}} ((\%iO_\mathrm{x})^* - (\%iO_\mathrm{x})^\mathrm{b}), \tag{12}$$

$$J_\mathrm{i} = \frac{\beta_\mathrm{g}\, \rho_\mathrm{g}}{M_{\mathrm{i,g}} R\, T} (P_\mathrm{i}^* - P_\mathrm{i}^\mathrm{b}), \tag{13}$$

where $\beta_\mathrm{m}$, $\beta_\mathrm{s}$ and $\beta_\mathrm{g}$ are the mass transfer coefficients for the metal, slag, and gas phase. $\rho_\mathrm{m}$ , $\rho_\mathrm{s}$ and $\rho_\mathrm{g}$ are the density of the metal, slag, and gas phase, $M_\mathrm{i}$, $M_{\mathrm{iO_x}}$, and $M_{\mathrm{i,g}}$, are the molar mass of species $i$ in metal, slag, and gas. $[\%i]^\mathrm{b}$, $[\%i]^*$, are the weight percentages of species $i$ in the metal bulk and at the interface, $(\%iO_\mathrm{x})^\mathrm{b}$ is the weight percentage of species $i$ in the slag bulk.

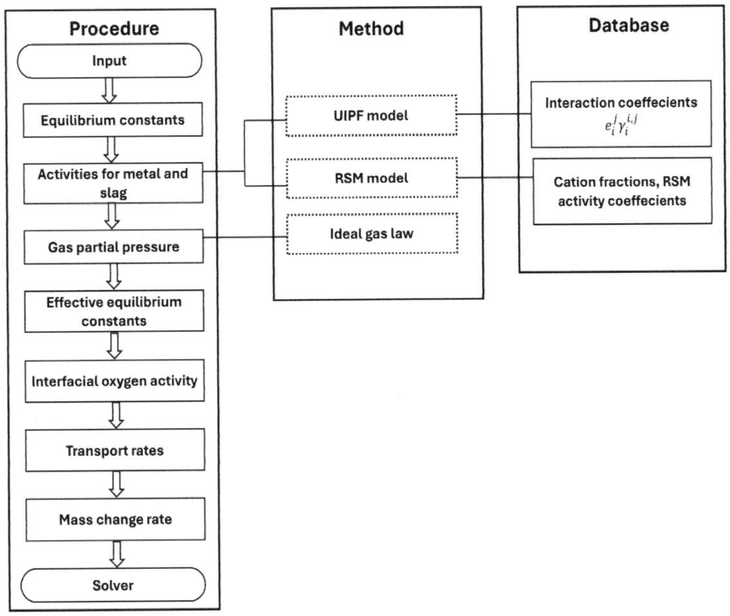


**Figure 1.** The calculation flow chart of the reaction model.

The model employs the explicit Runge–Kutta–Fehlberg (RKF–45) method for time integration, which incorporates an adaptive time–step size. This enables efficient and accurate tracking of composition changes over time by dynamically controlling the step size. The flow chart of the reaction model is shown in Figure 1.

## 3. Experimental Validation

An arc melting furnace with a volume capacity of 18 L was used for hematite reduction experiments reported by Putaala et al. [17]. A 15 g cylindrical pellet was prepared by weighing and compressing it to 20 mm in diameter and 15 mm in height. The pellet was placed in a copper crucible, and the chamber was sealed and vacuum–pumped. The chamber was then filled with a gas mixture of 90% argon and 10% hydrogen to a pressure of 1 atm, with a flow rate of 5 l/min, assuming a standard temperature and pressure (STP) state of gas (273.15 K and 1 atm). The plasma arc was ignited between a tungsten electrode and the copper crucible. The experiment was conducted in 2.5–minute intervals, with the chamber atmosphere being evacuated after each step to remove dust and reaction gases, followed by refilling with fresh gases. The total reduction time for the experiment was 12.5 minutes. The initial oxide composition is calculated based on the chemical analysis presented by Souza Filho et al. [18] and is given in Table 1.

**Table 1.** Chemical composition of the initial oxide phase

| Composition (wt%) | | | | | | | | |
|---|---|---|---|---|---|---|---|---|
| $Fe_2O_3$ | $SiO_2$ | $Al_2O_3$ | MnO | $TiO_2$ | CaO | MgO | S | $P_2O_5$ |
| 99.16 | 0.21 | 0.16 | 0.11 | 0.07 | 0.03 | 0.08 | 0.14 | 0.03 |

The initial values used to simulate the plasma–metal–slag reaction model are shown in Table 2.

**Table 2.** Initial values of different parameters used in this study.

| Parameter | Unit | Value |
| --- | --- | --- |
| Mass of the slag | g | 14.9 |
| Mass of the metal | g | 0.1 |
| Gas flow rate | l/min | 5 |
| Gas composition | vol % | 10% $H_2$-90% Ar |
| Temperature | °C | 2500 |
| Pressure | atm | 1 |
| $\rho_s$ | $kg \cdot m^{-3}$ | 3180 |
| $\rho_m$ | $kg \cdot m^{-3}$ | 7000 |
| $\rho_g$ | $kg \cdot m^{-3}$ | 0.147 |
| $A_{metal-slag}$ | $m^2$ | 0.000314 |
| $A_{melt-gas}$ | $m^2$ | 0.00001963 |
| $\beta_s$ | $m \cdot s^{-1}$ | 0.00005 |
| $\beta_m$ | $m \cdot s^{-1}$ | 0.00007 |
| $\beta_g$ | $m \cdot s^{-1}$ | 0.005 |
| $G_{H_2O}, G_i$ | $mol \cdot m^{-2} \cdot s^{-1}$ | 0.5 |

Based on the videos recorded during experiments, it was seen that the pellets completely melted in approximately 30 seconds. Therefore, in this study, the reduction occurring during the melting phase is not considered, as the melting of the hematite pellet is so fast under the hydrogen plasma atmosphere. The diameter of the reduced sample after reduction is used to calculate $A_{metal-slag}$, whereas the assumption that a 1/4th of the surface was in contact with the plasma arc is used to calculate $A_{melt-gas}$. Extremely high temperatures characterize the plasma zone. Optical emission spectroscopy (OES) tests under similar process conditions yielded temperatures of up to 10,000 K [19]. It is estimated that the temperature at the surface of the melt is about 1600–3100 °C [20]. In this study, the surface temperature of the melt is considered to be 2500 °C. The composition of plasma at 2500 °C was determined using a gas phase reaction module, based on the work of Jussila [21]. The basis of the model is formulated based on the Lagrangian approach for Gibbs energy minimization by White et al. [22]. Plasma species ($H_2$, H) were added in the gas phase reaction module, and the composition of the 10 vol% $H_2$ + 90 vol% Ar gas mixture was calculated. The model predicts that about 7 mol% of $H_2$ and 3 mol% of H are present in the system under equilibrium conditions for this temperature.

## 4. Results and discussion

The oxide phase composition during the reduction process as a function of time is shown in Figure 2. Only the major iron oxide species are considered, although it needs to be kept in mind that the ore also contains minor amounts of other oxides ($SiO_2$, $Al_2O_3$, MgO, and MnO). Initially, only hematite is present in the slag phase, it is reduced into magnetite and wüstite mixture. This gradual disappearance of $Fe_2O_3$ and $Fe_3O_4$ suggests that the reduction process effectively converts these higher oxidation states of iron into FeO. As the reduction continues, FeO becomes the dominant oxide in the slag, which is a clear indication that the hematite is being progressively reduced to wüstite and eventually to metallic iron. The slow reduction of hematite and magnetite could be because the conversion factors of activities for hematite and magnetite are not available, and this could lead to false activity calculation, and thus the slow reduction trend. How all the species collectively affect the reduction of hematite is still unknown and needs to be evaluated in more detail in later studies. This trend reflects the efficiency of the HPSR in progressively reducing iron oxides to metallic iron.

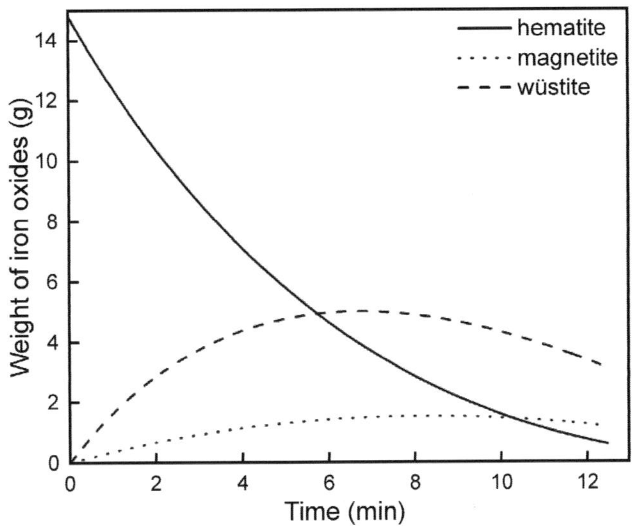


**Figure 2.** Weight of different iron oxides under hydrogen plasma.

Figure 3 displays the reduction of iron oxide to iron over time, showing a gradual increase in the weight of metallic iron as the reaction progresses. This suggests that the reduction of iron oxide occurs efficiently, with metallic iron being progressively formed. The model predicts a final iron weight of 6.79 g, about 3% higher than the experimental value of 6.62 g. In view of the various simplifications of the model, the agreement can be considered very good.

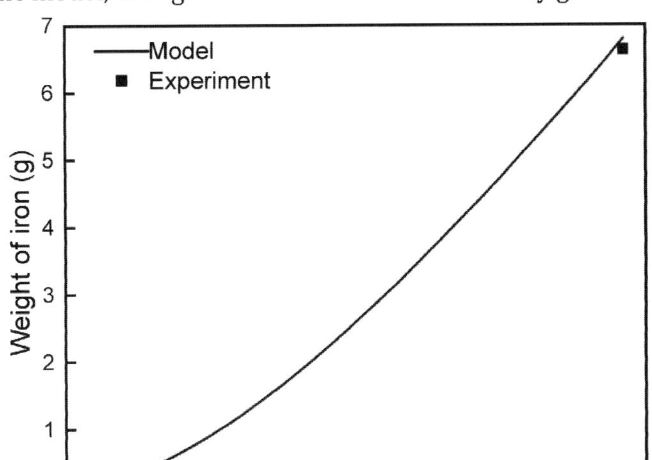


**Figure 3.** Weight of the iron as a function of the exposure time of the hydrogen plasma.

Figure 4 illustrates the metallization degree of iron over time, showing a steady increase in metallization as the process continues. The model predicts a final metallization degree of 67.2%, while the experimental results show a slightly lower final degree of 63.1%. Overall, the model demonstrates good agreement with experimental data, demonstrating the reduction of hematite to iron.

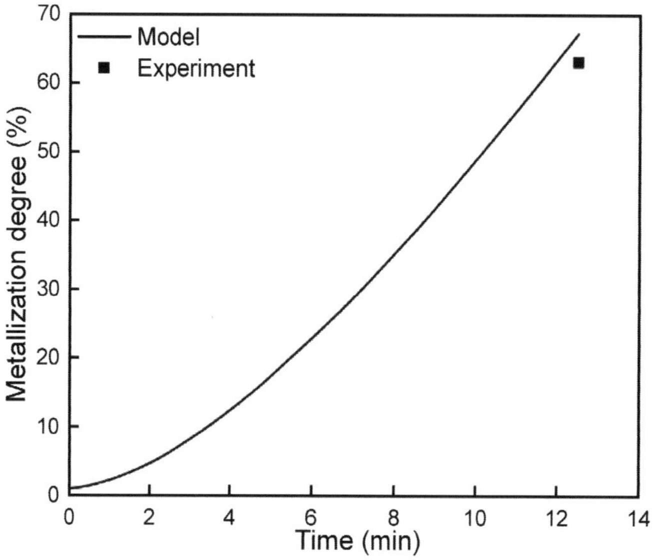


**Figure 4.** Effect of reduction time on iron metallization degree (%).

Figure 5 compares the model-predicted reduction kinetics of hematite and the experimental results reported by Souza Filho et al. [7] and Kannan et al.[11]. The experimental data by Souza Filho et al. [7] represent hematite reduction under a static gas atmosphere, where no continuous gas flow was maintained. These experiments were carried out at an initial pressure of 900 mbar, and the system remained closed throughout the reaction. Kannan et al. [11], shows the reduction of hematite powder, which is approximately 10 times faster than that of pellets under conditions like Souza Filho et al. [7]. In contrast, the model simulates the reduction process under a continuous gas flow, maintained for 12.5 minutes at 1 atm pressure. Despite the differences in experimental setup, static versus dynamic gas flow, the figure shows that the model successfully captures the general reduction trends observed in the experiments, offering insight into how different flow conditions influence reduction kinetics. The experimental data is shown for hematite samples with varying masses: 9 g (green circles), 15 g (red squares), powder hematite (blue triangles), while the solid black line represents the model's predictions.

The experimental data for a 15 g hematite pellet [7] shows an initial slow reduction rate. This slow initial reduction rate for 1–2 minutes is because hematite was converted into a mixture of magnetite and wüstite. After 2 minutes, the significant iron oxide in the system was wüstite, which is reduced to metallic iron, leading to an increase in the weight of the iron. At about ten minutes of reduction, the curves show a slow iron rise due to the sluggish kinetics of the wüstite. In contrast to the experimental results, the model predicts a slower reduction after 2 minutes. This could be due to the assumption that the total iron present at the interface between the metal and slag is contributed by all the iron oxides in the system, and the model does not accurately consider the reduction of the hematite and magnetite. This implies that the model may understate the variables influencing the rate of reduction in the experimental configuration. The model might also underestimate the effect of hematite and magnetite reduction due to the way the gas flow conditions are modeled. We plan to conduct further experiments and model under varying gas conditions to refine the model and better account for these factors.

While the model offers a reasonable approximation of the reduction kinetics, it slightly underestimates the reduction rate under the conditions tested. These differences suggest areas for model improvement, particularly in how it captures the influence of gas flow and other key variables. Future work will focus on testing different gas conditions to fine-tune the model for better accuracy.

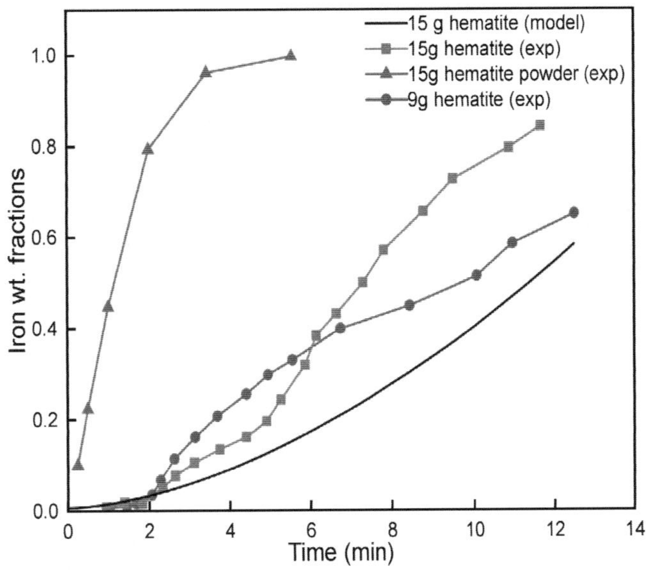


**Figure 5.** Iron conversion kinetics calculated in this study in comparison with 9 grams and 15 grams of hematite studied by Souza Filho et al. [7], and 15gms of hematite powder by Kannan et al. [11].

## 5. Conclusions

A comprehensive plasma–metal–slag kinetic model based on the EEC approach was developed to simulate the reduction behavior of hematite under hydrogen plasma conditions. While previous research shows kinetic interpretations of experimental HPSR data, there remains a need for a dynamic model that integrates mass and heat balance. The present model addresses this gap by adding gas–phase reactions to the metal–slag reactions. This enables us to understand the mass transfer–controlled competition between oxidation and reduction reactions at the metal-oxide interface. The EEC approach allows for the dynamic evaluation of the reaction kinetics, making it feasible to model the HPSR system's complex interactions.

In the experimental validation of the kinetic model for the reduction of hematite by HPSR, the model's predictions were compared with laboratory-scale experimental data. The experiments were conducted using a 15 g hematite pellet subjected to a hydrogen plasma under controlled conditions in an arc–melting furnace. The model predictions showed good agreement with experimental data for an overall reduction of hematite to metallic iron in a laboratory-scale HPSR furnace. The present model represents a significant advancement in understanding the kinetics of HPSR and serves as a robust framework for further model development towards a comprehensive process model. Future work should focus on refining model parameters by integrating actual charge material properties, incorporating the dynamic calculation of mass and heat transfer coefficients, and validating the stepwise pathway of iron oxides.

## Acknowledgements

The authors acknowledge the funding from Horizon Europe under the H2PlasmaRed project no.101138228.

## References

[1] World Steel Association. 2020 Steel's Contribution to A Low Carbon Future and Climate Resilient Societies—World Steel Position Paper; World Steel Association: Brussels, Belgium

[2] Heidari A, Niknahad N, Iljana M and Fabritius T 2021 A Review on the Kinetics of Iron Ore Reduction by Hydrogen *Materials* **14** 7540

[3] Nogami H, Kashiwaya Y, Yamada D 2012 Simulation of blast furnace operation with intensive hydrogen injection *ISIJ International* **52** 1523–7

[4] Spreitzer D, Schenk J 2019 Reduction of Iron Oxides with Hydrogen—A Review *Steel Res Int* **90**

[5] Seftejani M N, Schenk J 2018 Fundamentals of hydrogen plasma smelting reduction (HPSR) of iron oxides, a new generation of steelmaking processes Hydrogen Plasma Smelting Reduction (HPSR) of iron ore *Asia steel international conference*

[6] Sabat K C, Murphy A B 2017 Hydrogen Plasma Processing of Iron Ore *Metallurgical and Materials Transactions B: Process Metallurgy and Materials Processing* Science **48** 1561–94

[7] Souza Filho I R, Ma Y, Kulse M, Ponge D, Gault B, Springer H , Raabe D 2021 Sustainable steel through hydrogen plasma reduction of iron ore: Process, kinetics, microstructure, chemistry *Acta Mater* **213**

[8] Nagasaka T, Hino M, Ban-Ya S 2000 Interfacial kinetics of hydrogen with liquid slag containing iron oxide *Metallurgical and Materials Transactions B* **31** 945–55

[9] Zarl M A, Ernst D, Cejka J, Schenk J 2022 A New Methodological Approach to the Characterization of Optimal Charging Rates at the Hydrogen Plasma Smelting Reduction Process Part 1: Method *Materials* **15**

[10] Ernst D, Manzoor U, Souza Filho I R, Zarl M A, Schenk J 2023 Impact of Iron Ore Pre-Reduction Degree on the Hydrogen Plasma Smelting Reduction Process *Metals (Basel)* **13**

[11] Kannan R, Stevens A G, Fancher C, Rahman O, Miller R, Ward J, Ziabari A K, Babu S S, Nandwana P 2024 Reduction kinetics of hematite powder using argon/hydrogen plasma with prospects for near net shaping of sustainable iron *Sustainable Materials and Technologies* **40** e00879

[12] Mäkelä I, Visuri V-V, Aula M Model description for a dynamic mathematical EAF model: metal-slag-reactions (in preparation)

[13] Robertson D.G.C., Deo B, Ohguchi S 1984 Multicomponent mixed-transport-control theory for kinetics of coupled slag/metal and slag/metal/gas reactions: application to desulphurization of molten iron *Ironmaking and Steelmaking* **11** 41–55

[14] Ohguchi S, Robertson D.G.C., Deo B, Grieveson P 1984 Simultaneous dephosphorization and desulphurization of molten pig iron *Ironmaking and Steelmaking* **11** 202–213

[15] Ban-Ya S 1993 Mathematical Expression of Slag-Metal Reactions in Steelmaking Process by Quadratic Formalism Based on the Regular Solution Model *ISIJ International* **33** 2–11

[16] Elliot J F, Gleiser M, Ramakrishna V 1963 Thermochemistry for steelmaking: thermodynamic and transport properties **2**

[17] Putaala H-R, Pauna H, Javed A, Manzoor U, Klapproth D, R. Souza Filho I, Visuri V-V, Ganguly A, Huttula M, Fabritius T, Raabe D 2025 Effect of furnace parameters on optical emission spectra of hematite reduction by hydrogen plasma *Metallurgical and Materials Transactions B* accepted

[18] Souza Filho I R, Ma Y, Raabe D, Springer H 2023 Fundamentals of Green Steel Production: On the Role of Gas Pressure During Hydrogen Reduction of Iron Ores *JOM* **75** 2274–2286

[19] Pauna H, Ernst D, Zarl M, Aula M, Schenk J, Huttula M, Fabritius T 2022 Hydrogen plasma smelting reduction process monitoring with optical emission spectroscopy - Establishing the basis for the method *J Clean Prod* **372**

[20] Zulhan Z, Hakim B, Hendrawan Y, Hidayat T, Hermansyah E F, Yusro A H, Pambudi A L 2024 Rapid iron extraction from nickel leaching residue by a hydrogen plasma smelting reduction *Journal of Materials Research and Technology*, **30** 5346–5355

[21] R Jussila 2021 Modelling of gas-phase reactions in the freeboard of an EAF Master Thesis, University of Oulu

[22] White W B, Johnson S M, Dantzig G B 1958 Chemical Equilibrium in Complex Mixtures *J Chem Phys* **28** 751–755

Journal of Physics: Conference Series **3050** (2025) 012008 doi:10.1088/1742-6596/3050/1/012008

# Modeling the thermal properties of H-DRI during melting in an electric arc furnace

**A Agnihotri*, P Sulasalmi, and V-V Visuri**

Process Metallurgy Research Unit, University of Oulu, PO Box 4300, 90014 University of Oulu, Finland

*E-mail: ankur.agnihotri@oulu.fi

**Abstract.** Understanding the melting behaviour of hydrogen direct-reduced iron (H-DRI) is essential for further developing new ore-based steelmaking routes that utilize hydrogen reduction. This paper introduces a novel approach for deriving simple functions to predict the main thermochemical properties of DRI pellets, most notably the liquidus and solidus temperatures and enthalpy, which are critical for accurately predicting the energy balance of an electric arc furnace (EAF). The approach is demonstrated for a specific H-DRI grade. First, the thermochemical properties of the H-DRI are experimentally determined through differential scanning calorimetry (DSC) from room temperature up to 1600 °C. The experimental results are then compared with computational thermodynamic simulations using FactSage software. After validating the accuracy of the simulations, synthetic data is generated to derive regression equations for thermal properties using machine learning. The functions obtained from this regression analysis apply to EAF process models.

**Keywords:** Steelmaking, electric arc furnace, direct reduced iron, computational thermodynamics, differential scanning calorimetry, machine learning.

## 1. Introduction

EAF steelmaking has emerged as an essential route toward steel sector decarbonization. Today's traditionally scrap-based electric arc furnace (EAF) practice is transforming to include H-DRI, an eco-friendly feedstock subjected to reduction by hydrogen. The reason behind this transformation is the need to stabilize the quality of the feedstock, reduce dependency on varied scrap supplies, and lower $CO_2$ emissions (Hornby & Brooks, 2021)**[1]**. H-DRI has attracted significant academic attention because of its production method, which involves the reduction of iron ore by pure hydrogen, ultimately resulting in the formation of nearly pure Fe iron and water vapor instead of $CO_2$. This quality makes H-DRI an important technological advancement in the pursuit of near-zero carbon emissions in steel production**[2]**. Integrating H-DRI, however, brings fresh thermal and kinetic characteristics needing whole modeling to ensure proper EAF control. Relative to conventional scrap and DRI, H-DRI shows special features. Usually comprising 90–95% metallic iron, H-DRI also contains residual iron oxides, mainly FeO and minor gangue minerals like silica and alumina.

In contrast to carbon-containing DRI, where the in-situ oxidation of carbon can provide exothermic heat, H-DRI is nearly carbon-free. Geonu Kim et al. 2019**[3]** highlighted that the absence of carbon precludes beneficial oxidation reactions and instead results in the formation of an "autogenous slag" within individual DRI particles as the residual oxides melt. The presence of slag affects not only the efficiency of thermal conductivity in the particles but also the dynamics of heat transfer and the advancement of the melting front.

Studies in this field have applied different modeling techniques for the improvement of the effectiveness of EAF. Some of the techniques include thermodynamic energy balance models, kinetic melting models, and computational fluid dynamics (CFD) simulations.

Kirschen et al. (2021)**[4]** conducted a comprehensive study on process optimization for melting DRI in EAFs, focusing on slag operation. With the analysis of 16 EAF slag compositions and the development of a dynamic mass and energy model, the control of basicity of the slag and ensuring MgO saturation was shown to significantly reduce FeO losses and reduce electricity consumption. The models provided insightful observations on the influence of slag form and charging methods for DRI on furnace operating efficiency. Lin et al. (2024)**[5]** further went on to model packed beds of preheated H-DRI samples and show that inter-sample interactions and the presence of "icebergs" (low effective surface clusters) can further reduce the overall melt dynamics. The findings point to the intrinsic need for thermal conditions and particle structure for maximal melting efficiency of H-DRI in electric furnaces.

In complementing these approaches, CFD has become a useful tool for modeling the complex conditions that exist in an EAF. CFD simulations account for spatial variations in temperature, fluid flows, and electromagnetic effects due to arc heating and turbulence mixing. Ugarte et al. (2024)**[6]** utilized CFD to investigate the influence of layering the charge and of convective heat flux on melting performance and proved that local heat fluxes and turbulent conditions play a central part in determining the melting kinetics. Calderon Hurtado et al. (2024**[7]** Utilized combined computational and experimental work for the evaluation of the melting behaviour of pre-heated H-DRI in steel and slag. Contrary to earlier models that assumed constant thermophysical properties, their approach consisted of incorporating thermophysical property dependencies on temperature, including thermal conductivity, specific heat capacity, and density, thus addressing the implications of porosity and phase change. With COMSOL Multiphysics software, they developed a detailed heat transfer model that was validated through immersion experiments using embedded thermocouples. The research established that solid shell formation, convection heat transfer, and pellet porosity significantly contribute to determining melting kinetics. This hybrid model approach significantly improves modeling precision under conditions like those of an EAF. Notably, such simulations depend on accurate thermophysical property values (e.g., temperature-dependent specific heat, thermal conductivity, and latent heat), which are often obtained through DSC and similarity techniques.

Hybrid methods combining physics-based modeling with data-driven machine learning techniques have recently proven rather successful in accurately predicting the melting behaviour of H-DRI. Manojlović et al. (2022)**[8]** examined energy efficiency parameters in EAF operations using a data-centric machine learning approach. Using models such as support vector machines, random forests, XGBoost, and artificial neural networks, the analysis found important process variables, including oxygen and natural gas use, that strongly affect electrical energy need. The approach emphasized the importance of careful feature selection and data preprocessing to improve model performance. Such approaches are particularly valuable for H-DRI, where the complex interplay among residual oxides, autogenous slag formation, and convective heat transfer challenges conventional modeling methods. In the context of data-driven approaches, few studies have yet systematized machine learning for predicting thermophysical parameters of H-DRI, but the methodology aligns with recent advances in materials informatics.

By including DSC with FactSage databases into regression models, functions that link real data with computational forecasts are produced. This thereby marks a significant development that this study argues is a novel contribution. These derived equations are relevant not only for EAF process simulation but also apply to more comprehensive hybrid modeling systems meant to improve the optimization of sustainable steel production. Despite these developments, there does not exist an integrated framework that properly combines empirical evidence, thermodynamic modeling, and predictions made through machine learning. Most modern work focuses on individual techniques and does not entirely leverage the benefits potentially realized through the synergistic use of these methods.

By combining DSC-based thermal property measurements, FactSage simulations, and machine learning regression models to derive practical and regression functions for solidus temperature, liquidus temperature, and enthalpy of melting and specific heat, this study seeks to fill this gap. Based on H-DRI characteristics, these functions will be included in EAF digital twins and process models to allow predictive control and energy optimization according to H-DRI characteristics.

## 2. Materials and methods

### *2.1 Characteristics of H-DRI*

In this study, the samples of H-DRI were provided by an industry partner. The Centre for Material Analysis analyzed the H-DRI chemical composition using X-ray fluorescence (XRF). The results of XRF showed that the elemental composition was rather detailed and included major gangue elements along with metallic Fe content. The chemical composition of the H-DRI pellets studied is given in Table 1, while illustrative photographs are shown in Figure 1.

**Table 1.** Chemical composition of the H-DRI/wt %

| Fe | $SiO_2$ | CaO | $Al_2O_3$ | MnO | $P_2O_5$ | S | Reduction | Metallization, % | Size/mm |
|---|---|---|---|---|---|---|---|---|---|
| 93.65 | 3.44 | 1.74 | 0.78 | 0.23 | 0.12 | 0.04 | 100 %$H_2$ | 95 | 12*11*13 |

**Figure 1.** H-DRI pellet with 95% metallization.

The pellet exhibits a high metallic iron (Fe) ratio of 93.65 wt%, with minor amounts of oxides such as silicon dioxide ($SiO_2$, 3.44 wt%), calcium oxide (CaO, 1.74 wt%), aluminum oxide ($Al_2O_3$, 0.78 wt%), manganese oxide (MnO, 0.23 wt%), phosphorus pentoxide ($P_2O_5$, 0.12 wt%), and sulfur (S, 0.04 wt%). Considering the substantial metallization level (95%), the residual iron oxide (FeO) concentration was minimal and thus not assessed independently.

### *2.2 Computational thermodynamic modeling (FactSage)*

After the XRF characterization, computational thermodynamic modeling was undertaken utilizing FactSage (version 8.3). The information about composition derived from X-ray fluorescence (XRF) was directly used for simulations of equilibrium across a temperature continuum of 25°C to 1600°C at atmospheric pressure (1 atm). The thermodynamic databases employed encompassed FactPS (pure substances), FToxide (oxide systems), and FTmisc (miscellaneous multicomponent systems). These simulations generated predicted values for essential thermophysical properties, including liquidus and solidus temperatures, enthalpy, and specific heat capacity. These predictions can serve as reliable inputs to guide experimental design and further explain DSC results and study the phase transitions in the samples.

### *2.3 Analysis of differential scanning calorimeter*

Experimental validation of the computational predictions was performed using DSC on a Netzsch STA449F3 thermal analyzer. DSC experiments utilized the predicted FactSage thermophysical properties as initial guidance to design precise temperature programs and validate experimental results. During the experiments, inert conditions prevailed by passing argon gas through purification on hot titanium filaments to provide oxygen in only trace quantities. The heating took place from 28°C to a maximum of 1600°C at a heating rate of 10 °C/min. The sample was put in magnesium oxide (MgO) capsulated pans on alumina slip-plate supports at the bottom of the pan to provide uniform heating throughout. Continuous heat flow measurements were recorded, clearly capturing all thermal events (solid-state transitions, onset of melting, peak melting temperature). The DSC experiment on the samples was repeated three times to verify the reproducibility. DSC heat flow data were carefully processed via baseline correction techniques using Origin 2024 software to determine enthalpic variations correlated with melting transitions. This baseline correction included careful subtraction of instrumental drift and the baseline signal, making it easier to accurately determine experimental solidus and liquidus temperatures, enthalpy of fusion, and specific heat capacity values.

### *2.4 Regression modeling*

#### *2.4.1 Multiple linear regression modeling for predicting solidus and liquidus temperatures*

In this research study, a multiple linear regression (MLR) model **[9]** was employed to estimate H-DRI's solidus and liquidus temperatures predicated on its chemical composition. To develop the model, we created a synthetic dataset. This dataset captures the usual variations we see during the production of H-DRI. Furthermore, four representative H-DRI samples were analyzed utilizing X-ray fluorescence (XRF) to corroborate the synthetic data and to ensure that the model predictions are rigorously anchored in the actual operational conditions.

A synthetic dataset of 500 samples was formulated, with the composition of each sample diversified across the specified ranges to simulate H-DRI: 86–95 wt% Fe, 2–5 wt% $SiO_2$, 1–3 wt% CaO, 0.5–3 wt% FeO, 0.1–1.5 wt% $Al_2O_3$, and 0.05–0.5 wt% MnO.

For each synthetic data, the corresponding solidus and liquidus temperatures were computed utilizing FactSage software. The limited XRF measurements obtained from actual H-DRI samples were employed to validate that the synthetic data accurately represent authentic process conditions. The comprehensive dataset was randomly segmented into training (80%) and validation (20%) subsets, thereby establishing a robust framework for both model calibration and independent performance assessment. The following regression equations were proposed for the solidus $T_{sol}$ and liquidus temperatures $T_{liq}$ of the H-DRI:

$$T_{sol} = \alpha_{sol} + \sum_{i=1}^{6} \beta_{i,sol} \, (\text{wt-}\%_i) \tag{1}$$

$$T_{liq} = \alpha_{liq} + \sum_{i=1}^{6} \beta_{i,liq} \, (\text{wt-}\%_i) \tag{2}$$

where $\alpha_{sol}, \alpha_{liq}$ are the intercepts, $\beta_{i,sol}, \beta_{i,liq}$ denote the regression coefficients for the $i^{th}$ oxide component, and wt-$\%_i$ is the weight percentage of the $i$th component (Fe, $SiO_2$, CaO, FeO, $Al_2O_3$, and MnO)

The model's parameters were estimated using ordinary least squares. This method aims to reduce the total differences between the temperatures predicted by FactSage and those produced by the multiple linear regression models. Python was employed for the coding implementation, incorporating libraries like NumPy, Pandas, and SciKit-Learn. Three metrics—including the determination coefficient ($R^2$), mean squared error (MSE), and mean absolute error (MAE)—were used to evaluate the performance for training and validation datasets.

### *2.4.2 Prediction of specific heat and change in enthalpy using the Shomate equation*

To accurately outline the temperature-dependent thermodynamic properties of H-DRI in both its solid and liquid forms, the Shomate equation was applied to quantify the standard molar specific heat capacity ($C_p^0$) and the enthalpy ($H^0$) within the specified temperature interval**[10]**. For a designated phase (solid or liquid), the specific heat capacity is defined as:

$$C_p^0(T) = f_0 + f_1 T + f_2 T^2 + f_3 T^3 + f_4 T^{-2} \tag{3}$$

where $C_p^0(T)$ is the specific heat capacity at constant pressure in J/(g×K), T is the absolute temperature in Kelvin, and $f_0, f_1, f_2, f_3, f_4$ are empirical coefficients determined for a specific phase and temperature range.

Integrating equation (3) according to temperature, the following enthalpy equation is obtained:

$$\Delta H^0(T) = f_0 T + \frac{1}{2}(f_1 T^2) + \frac{1}{3}(f_2 T^3) + \frac{1}{4}(f_3 T^4) - (f_4 T^{-1}) + f_5 - f_6 \tag{4}$$

Here, $\Delta H(T)$ is expressed in J/g, representing the energy required to raise the temperature of one gram of H-DRI from $T_{ref}$ to T. This calculation is helpful in experimental analysis (such as DSC) and energy balance modeling. In this equation (3-4), the coefficients $f_0$ to $f_6$ for both the solid and liquid phases were initially derived from thermodynamic databases such as the NIST-JANAF Thermochemical Tables and the NIST Chemistry WebBook**[10]**. Though these databases yield dependable literature values, the chemistry linked to the H-DRI system required a thorough refinement of these coefficients. To attain an accurate depiction of the enthalpy and specific heat capacity for our samples, we conducted non-linear least squares fitting utilizing Python's SciPy library. The validity of the enhanced Shomate model was investigated through a comparative analysis of the predicted specific heat capacity and enthalpy values against those acquired from empirical measurements and FactSage. Again, the statistical metrics coefficient of determination ($R^2$), mean squared error (MSE), and mean absolute error (MAE) were used for evaluating the parameter fit.

## 3. Results and discussion

### *3.1 Experimental validation of thermodynamic simulations*

DSC data were thoroughly analyzed after the DSC experiment to extract important thermal properties. Baseline correction was a key step in this process, aimed at separating the thermal response of the sample from systemic effects. A linear baseline was established, selecting data points before and after the thermal event, and this baseline was subtracted from the original DSC signal to correct for any instrumental or environmental drift. The enthalpy of fusion for each sample was determined by integrating the area under the corrected DSC curve corresponding to the melting transition (see Figure 2 (b), and Figure 2 (c)). This integration provided the heat absorbed during the phase change, providing information about the energy requirements to melt the DRI samples.

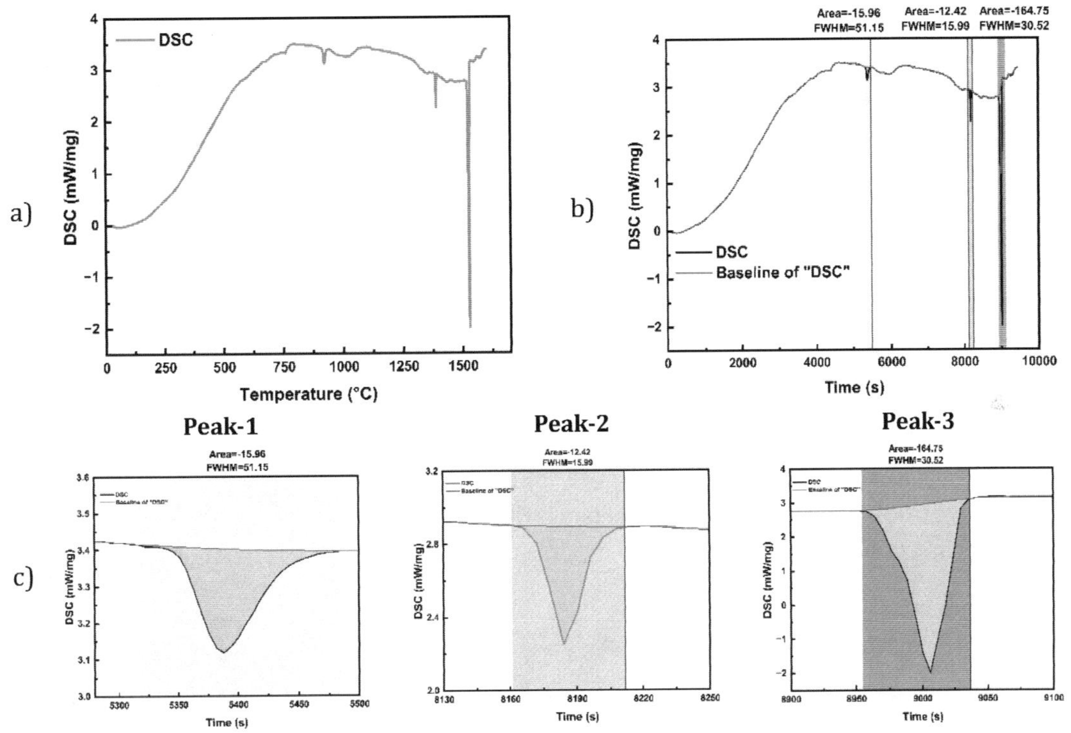


**Figure 2.** DSC analysis of H-DRI with 95% metallization:

a) DSC heat flow vs temperature, b) baseline-corrected DSC heat flow vs time, c) zoomed view of integrated thermal peaks

Table 2 compares the experimental enthalpy changes (ΔH) from DSC measurements with the theoretical predictions made using FactSage. For each thermal event, we calculated the experimental ΔH values by integrating the baseline-corrected DSC data over specific temperature ranges: 914–937 °C, 1386–1397 °C, and 1519–1532 °C. FactSage also provided predicted values for both ΔH and the related endothermic peak temperatures.

**Table 2:** Enthalpy change ΔH of H-DRI at different thermal events: experimental DSC measurements vs. Factsage calculations

| Thermal Event | $T_1$ (°C) | $T_2$ (°C) | ΔH (J/kg) | | % Difference |
|---|---|---|---|---|---|
| | | | DSC | FactSage | |
| Endothermic Peak 1 | 914 | 937 | −15.96 | −15.93 | 0.19% |
| Endothermic Peak 2 | 1386 | 1397 | −12.42 | −11.70 | 5.97% |
| Endothermic Peak 3 | 1519 | 1532 | −164.75 | −164.68 | 0.04% |

The DSC and FactSage results match closely for the first and third temperature ranges, with relative errors of 0.19% and 0.04%. However, the results for the temperature range of 1386–1397°C show a larger difference of about 5.97 %. This larger error likely comes from weaker signals in this region, making DSC measurements more sensitive to changes in the baseline and overlapping peaks. Additionally, the complex interactions between smaller phases in the system, like CaO, $Al_2O_3$, and $P_2O_5$, are not fully considered by the FactSage model's equilibrium assumptions.

### *3.1.1 Specific heat capacity $C_p$ of H-DRI: DSC measurements vs. FactSage predictions*

Specific heat capacity was determined using DSC data, with careful baseline correction designed to remove instrumental and environmental contributions to the raw heat flow data. Integration of heat flow over the relevant temperature intervals then allowed calculation of experimental $C_p$ values. Theoretical predictions of $C_p$ were obtained concurrently using FactSage software. Comparison of the values calculated using DSC with the FactSage predictions, together with percentage errors, provides insight into the accuracy of the thermodynamic model and the reliability of the experimental findings.

**Table 3:** Comparison of specific heat capacity $C_p$ of H-DRI at different temperatures: DSC experimental data vs. FactSage predictions

| T (°C) | $C_p$_(J/kg×K) | | % Difference |
|---|---|---|---|
| | DSC | FactSage | |
| 650 | 760 | 850 | 11.18 |
| 1100 | 880 | 980 | 10.75 |
| 1300 | 800 | 710 | 11.92 |

650°C, the value obtained from DSC was $C_p$ = 760 J/kg×K, while the value calculated by FactSage was 850 J/kg×K, leading to a difference of 11.18%. This difference can be mainly associated with the interactions between Fe and $SiO_2$ and the possible formation of FeO, which is calculated differently under equilibrium conditions in the FactSage model.

At 1100°C, the specific heat obtained from DSC was 880 J/kg×K while that from FactSage was 980 J/kg×K, showing a difference of 10.75%. In this temperature range, the process of transition between Fe and oxide components ($SiO_2$, CaO, $Al_2O_3$) to the liquid state becomes active. The FactSage calculations conducted assume an equilibrium state, while in actual DSC measurements, the imbalance and slow melting effects cannot be ignored, causing some discrepancy between the simulated and measured values.

At 1300°C, the highest error between DSC and FactSage was observed, where $C_p$ values were 800 J/kg×K and 710 J/kg×K, respectively. This 11.92% difference may be due to Fe–$SiO_2$ and FeO–$SiO_2$ interactions as well as the presence of $P_2O_5$ and MnO, which affect the melting and solubility at this temperature.

The FactSage software predicts the specific heat at a complete equilibrium state, while the DSC experiments reflect the actual physical and chemical changes, including dissociation processes among Fe, $SiO_2$, CaO, $Al_2O_3$, and $P_2O_5$. A difference of 10–12% was found between the $C_p$ values of DSC and FactSage over most temperature ranges, indicating that there is sufficient consistency between experimental and theoretical data, especially at higher temperatures. This study helps to understand the effect of chemical composition and temperature on the specific heat capacity, allowing accurate prediction of the thermal behavior of H-DRI.

### *3.2 Predicting solidus and liquidus temperatures using regression modeling*

The validity of the MLR model developed for estimating phase transition temperatures pertinent to H-DRI was rigorously examined through a synthetic dataset, which was also reinforced by limited experimental testing. We calibrated the model using 500 synthetic samples. We varied the chemical composition to match what is typically found in industrial H-DRI production. The composition included Fe (86–95 wt%), $SiO_2$ (2–5 wt%), CaO (1–3 wt%), FeO (0.5–3 wt%), $Al_2O_3$ (0.1–1.5 wt%, and MnO (0.05–0.5 wt%). Temperatures simulated by FactSage provided reference data, while selective XRF measurements ensured the fidelity of the synthetic data in mirroring actual process conditions. The regression equations were precisely formulated to represent the solidus ($T_{sol}$) and liquidus ($T_{liq}$) temperatures as linear relationships dependent on the compositional variables.

We used Python and the Scikit-learning package to train models, make predictions, and evaluate results. The best-fit regression equations for predicting the solidus and liquidus temperatures (in K) are given with intercepts set to $\alpha_{sol} = 1061.80$ K, $\alpha_{liq} = 1627.24$ K, based on the training dataset:

$$T_{sol}\ [K] = 1061.80 + 3.302 \times (\text{wt-\% Fe}) + 7.470 \times (\text{wt-\% SiO}_2 + 7.497 \times (\text{wt-\% CaO}) + 7.515 \times (\text{wt-\% FeO}) + 7.471 \times (\text{wt-\% Al}_2\text{O}_3) + 7.490 \times (\text{wt-\% MnO}) \quad (5)$$

$$T_{liq}\ [K] = 1627.24 + 1.629 \times (\text{wt-\% Fe}) + 1.613 \times (\text{wt-\% SiO}_2) + 1.571 \times (\text{wt-\% CaO}) + 1.585 \times (\text{wt-\% FeO}) + 1.571 \times (\text{wt-\% Al}_2\text{O}_3) + 1.802 \times (\text{wt-\% MnO}) \quad (6)$$

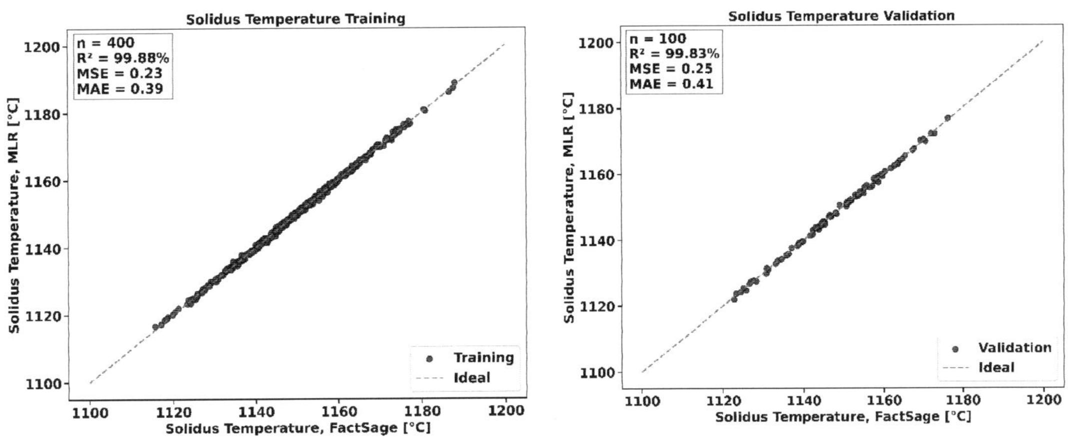


**Figure 3**. Comparison of measured and predicted solidus temperatures
(a) training dataset (left) and (b) validation dataset (right)

Figures 3(a) and 4(a) exhibit the scatter plots comparing the predicted temperatures against those simulated by FactSage for the training dataset, corresponding to $T_{sol}$ and $T_{liq}$, respectively. In both instances, the data points are closely clustered around the ideal y=x line, indicating a nearly perfect correlation and illustrating that the model effectively encapsulates the relationship between the phase transition temperatures and the chemical composition. Figures 3(b) and 4(b) provide analogous comparisons for the validation dataset. Despite a marginal increase in dispersion observed within these plots, the overall concordance remains exemplary, thereby affirming the model's capacity to generalize to new, previously unseen data.

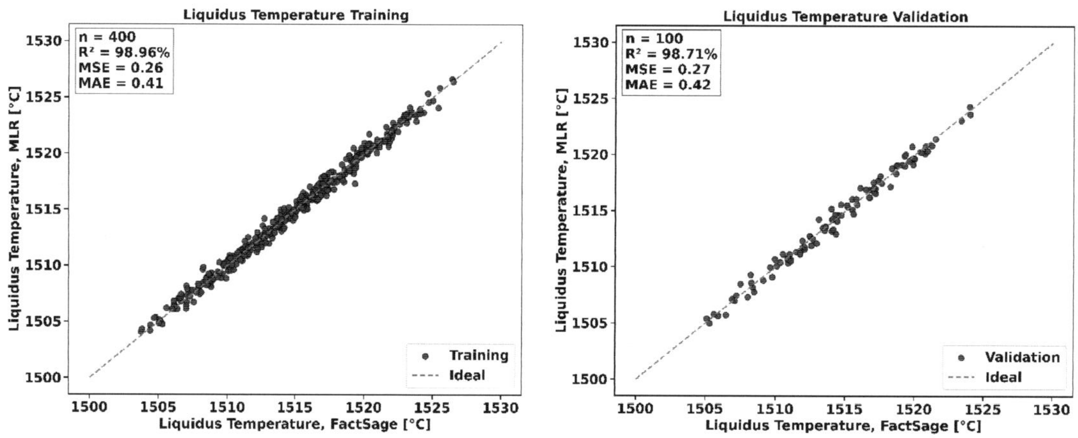


**Figure 4.** Comparison of measured and predicted liquidus temperatures (a) training dataset (left) and (b) validation dataset (right)

The robust correlation identified between the predicted and simulated temperatures substantiates the assertion that chemical composition is indeed the primary determinant influencing phase transition behavior in H-DRI. This level of predictive accuracy is essential not only for dependable energy balance calculations but also for the real-time monitoring and optimization of processes within industrial contexts.

### *3.3 Predicting specific heat and change in enthalpy using the Shomate equation*

Given the limited availability of experimental data, DSC measurements were conducted on a select subset of H-DRI samples. For the remaining samples, FACTSAGE simulations provided reference values for both specific heat and integrated enthalpy change. This dual approach allowed us to validate our regression model against experimental and simulated data. (see Figures 5 and 6).

Setting the intercept to $\alpha = 0$, the best fit for the predicted enthalpy and specific heat in the training dataset was obtained using the following regression equations:

$$C_{p,sol}\ [kJ/(kg \times K)] = 0.005920 \times (\text{wt-\% Fe}) + 0.007333 \times (\text{wt-\% } SiO_2) + 0.008687 \times (\text{wt-\% CaO}) + 0.004602 \times (\text{wt-\% FeO}) + 0.007888 \times (\text{wt-\% } Al_2O_3) + 0.006687 \times (\text{wt-\% MnO}) \quad (7)$$

$$H_{sol}\ [kJ/kg] = 10.6677 \times (\text{wt-\% Fe}) + 13.2275 \times (\text{wt-\% } SiO_2) + 16.0615 \times (\text{wt-\% CaO}) + 8.2922 \times (\text{wt-\% FeO}) + 14.5601 \times (\text{wt-\% } Al_2O_3) + 12.3304 \times (\text{wt-\% MnO}) \quad (8)$$

$$C_{p,liq}\ [kJ/(kg \times K)] = 0.007468 \times (\text{wt-\% Fe}) + 0.008322 \times (\text{wt-\% } SiO_2) + 0.009629 \times (\text{wt-\% CaO}) + 0.004175 \times (\text{wt-\% FeO}) + 0.009514 \times (\text{wt-\% } Al_2O_3) + 0.007753 \times (\text{wt-\% MnO}) \quad (9)$$

$$H_{liq}\ [kJ/kg] = 12.3271 \times (\text{wt-\% Fe}) + 15.5888 \times (\text{wt-\% } SiO_2) + 18.0368 \times (\text{wt-\% CaO}) + 7.8211 \times (\text{wt-\% FeO}) + 17.8203 \times (\text{wt-\% } Al_2O_3) + 14.5226 \times (\text{wt-\% MnO}) \quad (10)$$

Similarly to equations 5 and 6, Equations (7-10) are valid for a six-component system containing Fe (86–95 wt% %), $SiO_2$ (2–5 wt %), CaO (1–3 wt %), FeO (0.5–3 wt %), $Al_2O_3$ (0.1–1.5 wt %), and MnO (0.05–0.5 wt %).

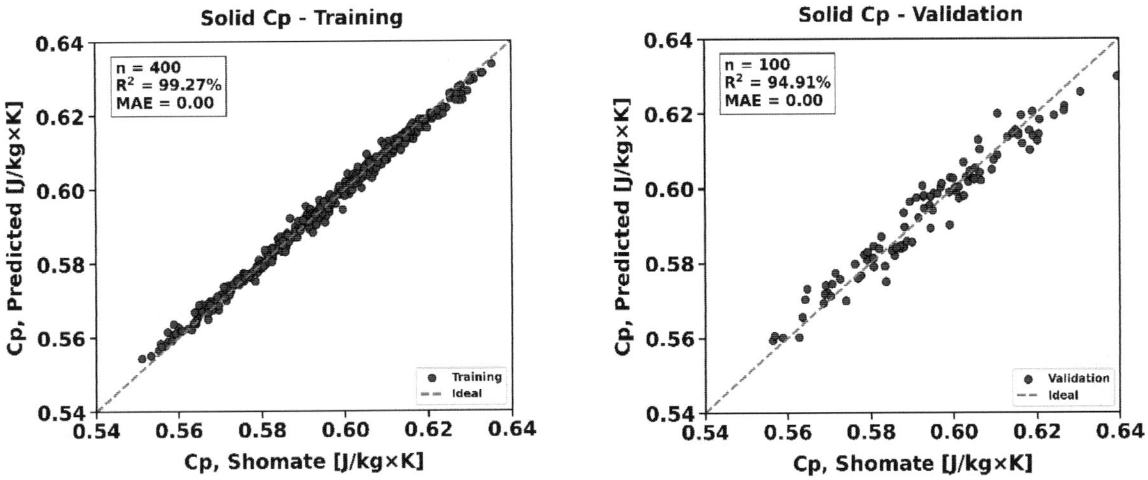


**Figure 5.** Comparison of measured and predicted solid phase heat capacity (a) training dataset (left) and (b) validation dataset (right)

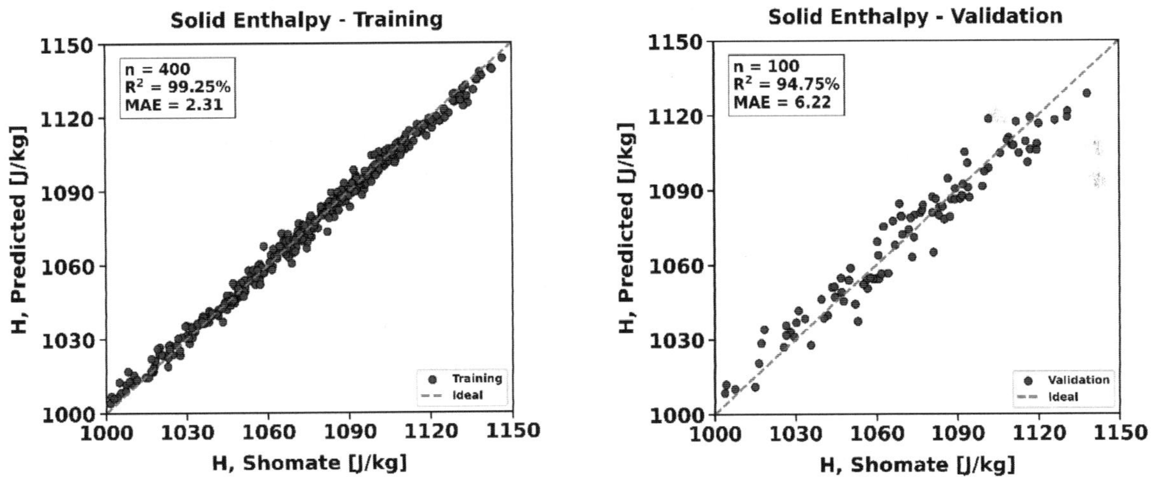


**Figure 6.** Comparison of measured and predicted solid phase enthalpy (a) training dataset (left) and (b) validation dataset (right)

For the solid phase, the regression model was initially calibrated employing a training dataset that encompassed synthetic data generated from FactSage simulations and DSC measurements derived from four representative samples. Figure 5 (a) presents the predicted specific heat capacity, $C_p(T)$, as a function of temperature for the solid phase obtained from the training dataset. The model effectively captures the predicted temperature dependence of the specific heat capacity, presenting a smooth curve that closely aligns with the experimental trend. The elevated coefficient of determination ($R^2 \approx 0.9927$) and minimal error metrics (mean squared error and mean absolute error) signify that the model accurately reproduces the inherent thermal behavior with considerable precision. Expanding upon these $C_p(T)$ predictions, numerical integration was executed across the temperature range to compute the enthalpy change, $\Delta H(T)$. Figure 6(a) depicts the integrated $\Delta H(T)$ for the solid phase derived from the training dataset.

The resultant enthalpy curve demonstrates a continuous increase with temperature, exhibiting excellent quantitative agreement with the FactSage simulation data. The negligible discrepancies observed between the predicted and reference ΔH values, as evidenced by the low mean squared error and mean absolute error values, further substantiate the validity of the integration methodology. The performance of the model was subsequently assessed utilizing an independent validation dataset. Figure 5(b) illustrates the predicted $C_p$(T) for the solid phase derived from the validation set. While the data points reveal a somewhat greater variability than the training dataset, the overall trend continues to be consistent, thereby indicating the model's robust generalizability. Figure 6(b) subsequently showcases the integrated ΔH(T) for the solid phase from the validation dataset. Despite minor variability, the predicted enthalpy changes align closely with the FactSage-simulated values, affirming the reliability of the model when applied to previously unseen data.

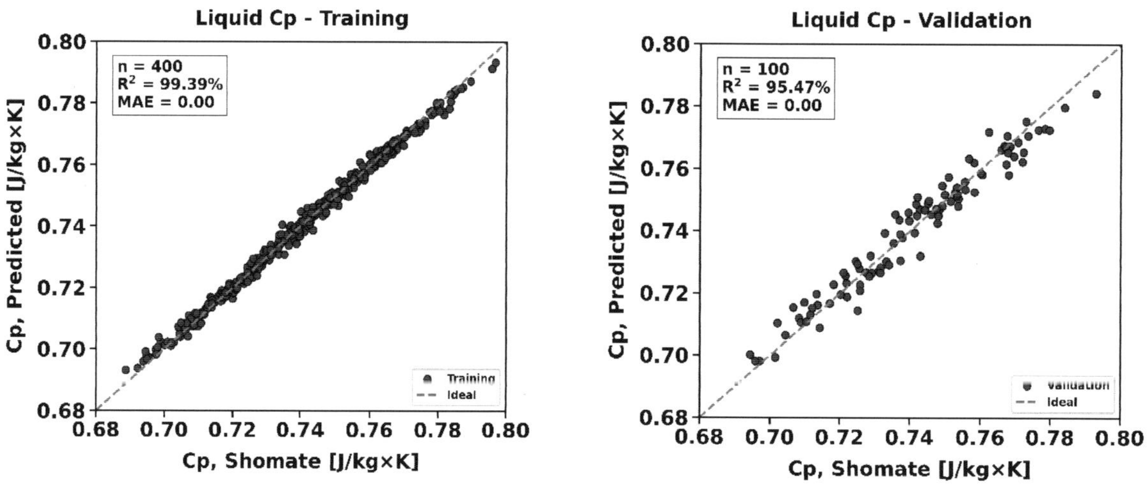


**Figure 7.** Comparison of measured and predicted liquid phase heat capacity (a) training dataset (left) and (b) validation dataset (right)

Similar analyses were conducted for the liquid phase. Figure 7(a) presents the predicted $C_p$(T) for the liquid phase derived from the training dataset, wherein the regression model captures well both the curvature and magnitude of the thermal response exhibited by the liquid phase. The integration of these data, as illustrated in Figure 8(a), yields $\Delta H$(T) values that are in strong concordance with the simulation results, with performance metrics indicating $R^2$ values are 0.9931 and exceedingly low error margins.

For the validation of the liquid phase, Figure 7(b) depicts the predicted Cp(T) for the liquid phase within the validation dataset. The specific heat capacity curve, while demonstrating slightly enhanced variability, remains favorable with the training profile, thereby reinforcing the robustness of the model. Figure 8(b) presents the integrated ΔH(T) for the liquid phase from the validation dataset. The close alignment observed between the predicted enthalpy changes and the FactSage reference data confirms that the model is proficient in accurately predicting the thermal energy requirements, even for independent samples.

Journal of Physics: Conference Series **3050** (2025) 012008 doi:10.1088/1742-6596/3050/1/012008

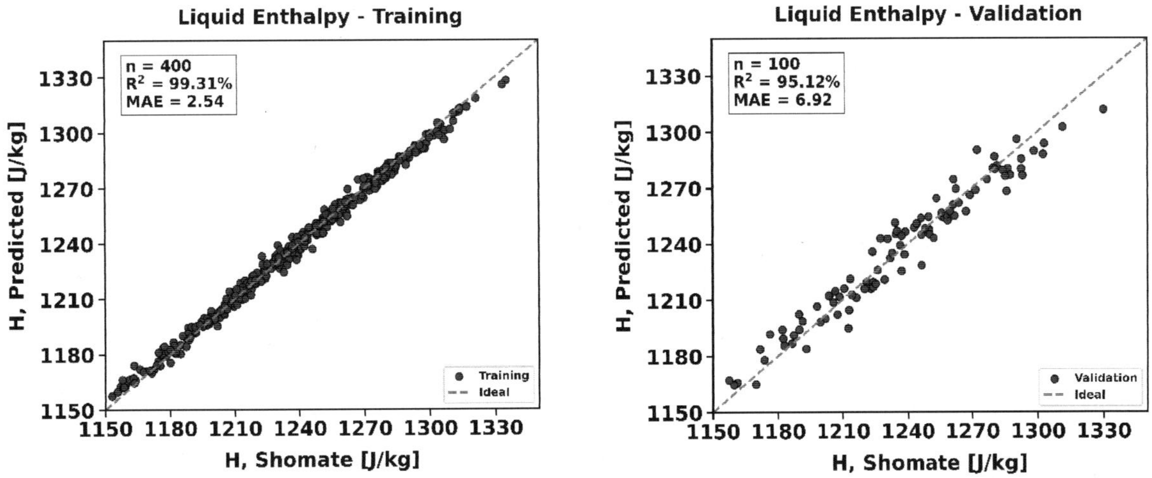


**Figure 8.** Comparison of measured and predicted liquid phase enthalpy
(a) training dataset (left) and (b) validation dataset (right)

In the current work, a machine learning model based on Shomate regression has been developed to predict the melting behavior of H-DRI compositions. The model was created with a synthetically generated dataset from FactSage that was subsequently validated with DSC measurements. It successfully learned the nonlinear enthalpy-temperature and specific-heat-temperature relationships, especially around the solidus and liquidus boundaries. The combination of advanced computational thermodynamics with accurate thermal simulations highlights the model's strength for real-time energy balance calculations in low-carbon EAF operations.

## 4. Conclusions

The aim of this was to develop a methodology for predicting the thermal properties of H-DRI and to derive regression equations to be used in EAF process models. To this end, DSC experiments were conducted to study the thermal properties of a selected H-DRI grade and to provide experimental data for validating how accurately the thermal properties can be simulated using the FactSage computational thermodynamics software. Finally, regression equations were derived for the specific heat capacity and phase change temperatures using a machine learning approach.

The key findings are as follows:

1. The high level of metallization in H-DRI, which is approximately 95% iron, along with its minimal oxide content, contributes to lower melting ranges. Analyses using DSC and FactSage indicate that the solidus and liquidus temperature ranges are between 1120-1170 °C and 1510-1530 °C, respectively. The precise measurement of phase transition temperatures indicates that the presence of minimal oxide impurities significantly affects thermal properties and energy balance calculations. These factors are crucial for efficient low-carbon steel fabrication.
2. The MLR model, validated using FactSage simulations, accurately predicts H-DRI phase transition temperatures, achieving an MAE of 0.41 and $R^2$ = 99.83 for the solidus and an MAE of 0.42 and $R^2$ = 98.71 for the liquidus.
3. At temperatures of 650 °C, 1100 °C, and 1300 °C, the specific heat capacities measured by DSC are 760, 880, and 800 J/(kg×K). These values are close to the predictions made by FactSage, which are 850, 980, and 710 J/(kg×K). The differences between the two sets of

values are 11.18%, 10.75%, and 11.92%, respectively.

4. The enthalpy values measured in the process of melting have insignificant differences for endothermic events. For the temperature range of 914–937 °C, the difference is quantified at 0.19%. For 1386–1397 °C, the variation is up to 5.97%, while for the range of 1519–1532 °C, the difference is just 0.04%. These observations validate the consistency of the computational method in calculating the enthalpy of fusion.
5. The experiments with DSC verified computational findings and generated accurate enthalpy values by measurement of baseline-corrected heat flow. The close correspondence between data from experiments and simulation confirmed the hybrid strategy's validity.
6. Shomate-based regression was more effective than normal linear and polynomial regression in representing the complex temperature behavior of H-DRI. It provided accurate, high-fidelity thermodynamic property predictions and proved to possess great potential as a process model instrument.

The study confirms that combining computational thermodynamics with advanced machine learning methods and high-precision thermal analysis creates an effective approach for predicting the melting behaviors of complex feedstocks, such as H-DRI. This hybrid method enhances the accuracy of energy balance calculations and allows for real-time optimization of low-carbon electric arc furnace operations. As a result, it represents a significant advancement toward clean steel production

## 5. Acknowledgements

This research was supported by the project "Future EAF steelmaking with different iron-bearing raw materials", funded by the European Union under the Helios-Horizon-MSCA-DN program (Project Number: 101120068). The work by V.-V. Visuri was supported by Towards $CO_2$-lean electric arc furnace steelmaking through fundamental and data-driven mathematical modeling (LEAF) project funded by the Research Council of Finland (no. 356439).

## 6. References

[1] Dr. Sara Hornby and Professor Geoff Brooks 2021 Impact of hydrogen DRI on EAF steelmaking: MIDREX report-second quarter
[2] Liu W, Zuo H, Wang J, Xue Q, Ren B and Yang F 2021 The production and application of hydrogen in steel industry International Journal of Hydrogen Energy **46** 10548–69
[3] Kim G, Kacar Y and Pistorius P C 2019 Carbon bonding state has a small effect on melting of Direct-Reduced Iron Metall Mater Trans B **50** 2508–16
[4] Kirschen M, Hay T and Echterhof T 2021 Process improvements for direct reduced iron melting in the electric arc furnace with emphasis on slag operation Processes **9** 402
[5] Lin X, Ni B and Shangguan F 2024 Numerical simulation of the hydrogen-based directly reduced iron melting process Processes **12** 537
[6] Ugarte O, Li J, Haeberle J, Frasz T, Okosun T and Zhou C Q 2024 CFD modeling of HBI/scrap melting in industrial EAF and the impact of charge layering on melting performance Materials **17** 5139
[7] Calderon Hurtado F A, Govro J, Emdadi A and O'Malley R J 2024 The melting behavior of hydrogen direct reduced iron in molten steel and slag: an integrated computational and experimental study Metals **14** 821
[8] Manojlović V, Kamberović Ž, Korać M and Dotlić M 2022 Machine learning analysis of electric arc furnace process for the evaluation of energy efficiency parameters Applied Energy **307** 118209
[9] Douglas C. M, Elizabeth A. P, and G. Geoffrey V. Introduction to linear regression analysis (John Wiley & Sons, 2012)
[10] Chase Jr. and M.W. (1998) NIST-JANAF Thermochemical Tables Journal of Physical and Chemical Reference Data, Monograph 9, 1-1951

Journal of Physics: Conference Series **3050** (2025) 012009 doi:10.1088/1742-6596/3050/1/012009

# Optimization Based Experimental Design of Metal–Slag Experiments in Hydrogen Plasma Smelting Reduction Process

**T Vuolio[1*], M Zarl[2], I Lappeteläinen[1], A Javed[1] and V-V Visuri[1]**

[1] Process Metallurgy Research Unit, University of Oulu, PO Box 4300, 90014 University of Oulu, Finland.
[2] Area 2, K1–MET GmbH, Stahlstraße 14, 4020 Linz, Austria

*E–mail: tero.vuolio@oulu.fi

**Abstract.** This study presents an optimization based experimental design scheme for design of metal–slag experiments in hydrogen plasma smelting reduction. The approach combines computational thermodynamics and machine learning techniques by making use of machine learning -based surrogate–modelling. The objective function for the optimization is formulated as a maximization problem, in which the conversion rate or reduction degree of iron oxide and the phosphorus partition ratio are treated as the main objectives. The objective function is maximized with respect to slag composition in a $CaO–SiO_2–FeO–Al_2O_3–MgO$ system. Additional constraints are introduced for maximum effective viscosity and slag/metal mass ratio to obtain results that are relevant for industrial operation.

## 1. Introduction

Steel is an integral part of today's society and industrial manufacturing. Over 1,660 million tons of iron and steel are produced annually. On the contrary, steelmaking accounts for up to 6.5–8% of global $CO_2$ emissions. However, the steel industry has managed to reduce energy consumption per ton of steel by 60% over the past 50 years [1], with the goal of reducing $CO_2$ emissions by 80% before the year 2050. For this reason, alternative technologies to blast furnace are needed for the reduction of iron ore [2, 3].

One of the promising technologies for reducing $CO_2$ emissions is Hydrogen Plasma Smelting Reduction (HPSR). In the HPSR process, hydrogen assumedly exists in a plasma phase, which reduces iron oxides into metallic iron [1]. The process offers possibility to simultaneously melt and reduce the iron ore, with $H_2O$ as the only end–product of the reduction. Several studies have been conducted using hydrogen plasma to produce steel, see for example Refs. [1–5], as well as on the hydrogen plasma characterization [6]. Detailed literature reviews on the process have already been provided [1,5]. However, the literature concerning the dephosphorization via slag–metal reaction is scarce [7], and the effect of slag has been considered only in the viewpoint of reduction efficiency [8].

The objective of this study is to present a mathematical tool for design of experiments that allow the selection of optimal slag composition in conditions relevant to the HPSR process. The tool combines the use of computational thermodynamics, machine learning and multi–objective optimization. The bearing thought in the framework is to significantly reduce the computational cost of the optimization phase by making use of machine learning -based surrogate models. The

Journal of Physics: Conference Series **3050** (2025) 012009 doi:10.1088/1742-6596/3050/1/012009

study presents a single design case with to some extent arbitrary, but realistic design objectives and briefly discusses the effects on the design constraints, fundamentals of dephosphorization and model selection on the optimization results.

## 2. Materials and methods

The considered model is a simplification of the dephosphorization stage in conditions relevant to the HPSR process. The HPSR process scheme is here divided roughly into reduction and dephosphorization stages. Thus, for example the pre-melting phase and all the alloying phases are intentionally ignored, which makes the computational model here a simple slag-metal - equilibrium system. The equilibrium content of phosphorus in the metal phase is given by

$$[\mathrm{P}]_{\mathrm{eq}} = \frac{[\mathrm{P}]_0 m_{\mathrm{metal}} + (\mathrm{P})_0 m_{\mathrm{slag}}}{L_{\mathrm{P}} m_{\mathrm{slag}} + m_{\mathrm{metal}}}, \tag{1}$$

where $[\mathrm{P}]_0$ is the initial content of phosphorus in the metal phase, $(\mathrm{P})_0$ is the initial content of phosphorus in the slag phase, $[\mathrm{P}]_{\mathrm{eq}}$ is the equilibrium content of phosphorus in the metal phase, $L_{\mathrm{P}}$ is the phosphorus partition ratio, $m_{\mathrm{metal}}$ is the mass of the metal phase, and $m_{\mathrm{slag}}$ is the mass of the slag phase.

The main compositional variables considered to affect the dephosphorization are the temperature, and CaO, $SiO_2$ and FeO contents of the slag. The metal–slag system model here is described as a steady-state system where the slag and the metal phases are in apparent thermodynamic equilibrium state. The phase compositions are computed using simple static mass–balance model. In other words, the dynamic behavior of the reduction process, slag formation and the dephosphorization, and the interactions of the phenomena is not considered because of the already high computational complexity of the system. Instead, the objective is to provide a computationally feasible experimental design scheme that allows the evaluation of multiple design alternatives in a multi component system. In the simulation and the optimization, the following main assumptions and simplifications are made

1) Only the last step of the iron oxide reduction (FeO → Fe) is considered, and consequently the iron exists either as Fe or FeO The residual FeO is considered to end up in the slag phase. The system properties do not affect the reduction degree l. conversion rate, but the slag composition is a consequence of the reduction.
2) The kinetics of the process, e.g., oxide dissolution, reduction rate and desphorization rate, are not accounted for.
3) The slag phase is either a heterogenous solution of spherical solid particles and liquid slag or a fully liquid homogenous solution
4) The system is isothermal, and consequently the optimal slag composition is valid only in the given conditions

The computational complexity in the optimization is reduced by using a surrogate-model approach, which combines computational thermodynamics and machine learning. The computational framework was implemented with Matlab R2020b. In practice, a full factorial

Journal of Physics: Conference Series **3050** (2025) 012009 doi:10.1088/1742-6596/3050/1/012009

design of computer experiments is used to construct a representative dataset with FactSage ver. 8.3 to describe dephosphorization capability and physical properties of the slag with respect to its composition. The data is then used as training data for a machine learning models (e.g. shallow regression neural networks), which allow a fast evaluation of the slag properties of a given composition. Herein, the computational complexity of optimization is significantly reduced, given that the objective function is evaluated thousands (population size x optimization rounds = $n_{pop}$ x $iter_{max}$) of times during the optimization. For example, with a genetic algorithm population size of 100 individuals and 100 optimization rounds, the objective function is evaluated 10,000 times. The workflow of the design tool is presented in **Figure 1**.

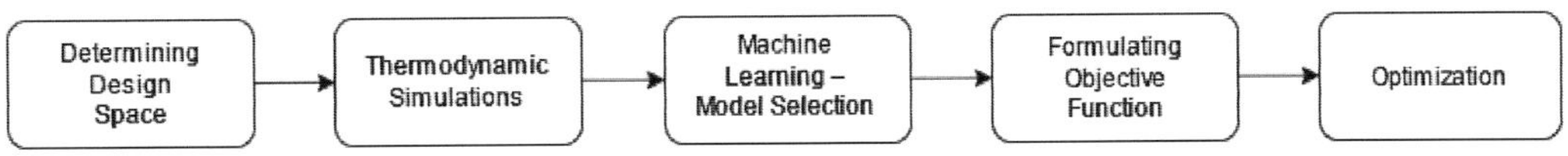


**Figure 1.** Workflow of the proposed framework.

*2.1 The objective function*

The objective of the optimization is to find a slag composition that simultaneously maximizes the reduction degree but also minimizes the phosphorus content in the metal phase. These objects are known to be conflicting, as the phosphorus partition ratio between the slag and the metal phase is known to be positively correlated with the FeO content of the slag, and the activity of the oxygen in the metal phase. The effective viscosity of the slag is given here as a constraint with an arbitrary value dependent on the design objective. The maximum slag-metal ratio is set here as 0.1. The objective function for the optimization is formulated as a maximization problem, with the reduction degree as the numerator and the phosphorus content as the denominator. The objective function is maximized with respect to slag composition in a CaO–$SiO_2$–FeO–$Al_2O_3$–MgO system. Additional constraints are introduced for maximum effective viscosity and slag/metal mass ratio to obtain results that are relevant for industrial operation. As stated above, the objective function values are evaluated using the surrogate models values. Thus, it is apparent that the objective function is complex and non-linear and thus for example gradient-based methods would be ill-suited for solving the task.

*2.2 Simulations for phosphorus partitioning between metal and slag*

The partitioning of phosphorus between liquid iron and HPSR slag under equilibrium conditions was studied using the equilib module of FactSage ver. 8.3 computational thermodynamics software. The considered liquid and solid solution phases include those from the FToxid and FSstel databases, as well as pure solid species from the FactPS database. The studied systems consisted of 1000 g of P-containing iron (99.9 wt-% Fe + 0.1 wt-% P) in equilibrium with 100 g of slag of various compositions under a pressure of 1 atm (101,325 Pa) at three temperatures: 1600 °C (1873.15 K), 1700 °C (1973.15 K), and 1800 °C (2073.15 K). The studied slag compositions follow the experimental design shown in Table 1. The experimental design matrix was constructed using principles of the full-factorial design. The calculations covered 5,760 compositions per temperature, resulting in a total of 17,280 simulation cases across the three temperatures. It can be easily seen that the accurate estimation of the effect of temperature in the range of 1600–1800 °C demands a rather dense computational space (e.g. step size roughly 1–5

Journal of Physics: Conference Series **3050** (2025) 012009 doi:10.1088/1742-6596/3050/1/012009

°C/K), and consequently 230,400–1,152,000 simulations. Thus, describing the effect of temperature is intentionally left as directional.

**Table 1.** Studied slag compositions.

| | Composition [wt–%] | | | | |
|---|---|---|---|---|---|
| Value | $SiO_2$ | $Al_2O_3$ | CaO | MgO | FeO |
| Minimum | 16 | 0 | 32 | 0 | 0 |
| Maximum | 60 | 10 | 60 | 10 | 30 |
| Step | 2 | 2 | 2 | 2 | 2 |

*2.3 Simulations for the viscosity of slag*

The workflow for calculating the effective viscosity of HPSR slag in equilibrium with P-containing iron is shown in Figure 2. The first step of the workflow involves the equilibrium calculations described in Subsection 2.2 and is not repeated here. The second step entails calculating the dynamic viscosity of the liquid portion of the slag. Finally, the third and fourth steps involve calculating the relative viscosity and the effective viscosity, respectively. A more detailed explanation of these four steps is provided in the following.

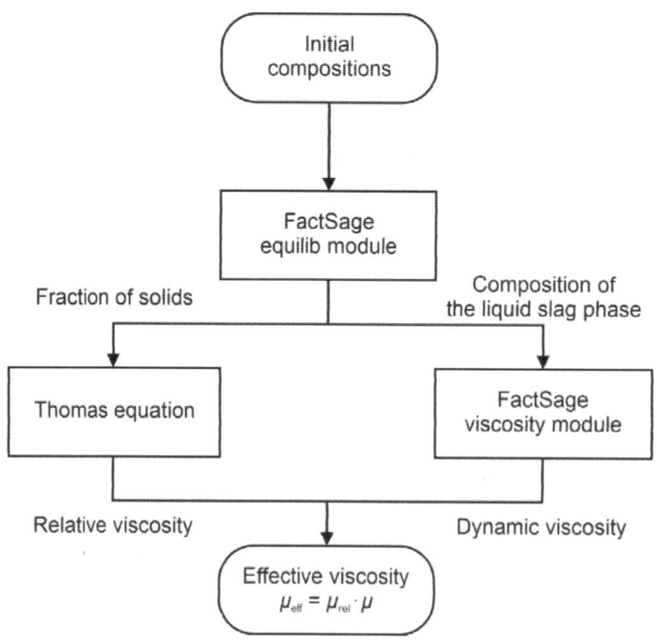


**Figure 2.** Workflow of the viscosity calculations.

The dynamic viscosity of the liquid slag phase was calculated using the *viscosity* module embedded in the FactSage ver 8.3 software. The calculations were performed using the *melts* viscosity database, which is valid for liquid and supercooled slags at high temperatures. To this end, the compositions of the liquid slag phase representing the $SiO_2$–$Al_2O_3$–CaO–MgO–FeO–$Fe_2O_3$–$P_2O_5$ system were extracted from equilibrium calculations (see Subsection 2.2) and normalized to a system size of 100 g, excluding $P_2O_5$ which is not included the *viscosity* module. Furthermore, the viscosity calculations for the studied $SiO_2$–$Al_2O_3$–CaO–MgO–FeO–$Fe_2O_3$ systems were performed only when a liquid slag phase was present. As a result, 56 cases were excluded, reducing the total number of viscosity calculations from 17,280 to 17,224. The results were stored as ln $\mu$ for accuracy but converted to $\mu$ in Pa·s. In practice, many of the studied slag compositions exhibit a noticeable fraction of solid phases. To address this, the viscosity of the liquid–solid mixture was described using the concept of *effective viscosity*:

$$\mu_{\text{eff}} = \mu_{\text{rel}} \cdot \mu, \tag{2}$$

where $\mu_{\text{rel}}$ is the relative viscosity and $\mu$ is the dynamic viscosity – in this case, the dynamic viscosity of the liquid slag. The shape of the solid particles is known to affect the relative viscosity, with spherical particles having the least impact. In this work, the solid particles were assumed to be spherical to provide a lower limit for the effective viscosity. The relative viscosity of the slag was calculated using the effective viscosity correlation proposed by Thomas [9], which assumes spherical particles. The correlation is valid for a solid fraction of $\phi_{\text{solids}} \leq 0.6$ (i.e., up to 60% by volume) and can be expressed as follows:

$$\mu_{\text{rel}} = 1 + 2.5\phi_{\text{solids}} + 10.05\phi_{\text{solids}}^2 + A\exp(B\phi_{\text{solids}}), \tag{3}$$

where $\phi_{\text{solids}}$ is the volume fraction of solids, while $A$ and $B$ are coefficients, with values of $A$ = 0.00273 and $B$ = 16.6. It is noteworthy that Eq. 3 has the following limit:

$$\lim_{\phi_{\text{solids}}\to 0} \mu_{\text{rel}} = 1.00273, \tag{4}$$

which indicates that using Eq. 3 introduces a relative error of 0.273% for completely molten slags. However, this error is negligible compared to the prediction errors of equilibrium and dynamic viscosity calculations. Due to the lack of suitable density data, it was assumed that the volume fraction of the liquid phase corresponds to its weight fraction. Consequently, the volume fraction of solids was calculated simply as

$$\phi_{\text{solids}} = 1 - \phi_{\text{liquid}} = 1 - y_{\text{liquid}}, \tag{5}$$

where $\phi_{\text{liquid}}$ and $y_{\text{liquid}}$ are the volume and weight fractions of the liquid slag phase, respectively. The weight fraction of the liquid phase was obtained from the equilibrium calculations described in Subsection 2.2.

### *2.4 The surrogate-models and objective function evaluation*

Separate models are trained for the phosphorus partition ratio, dynamic viscosity and liquid fraction of the slag phase. The dynamic viscosity and the liquid fraction models are implemented as shallow neural networks, whereas the phosphorus partition model is an ensemble of shallow neural network, a multivariate regression model trained and validated against literature data and several simple regression models acquired from the literature. The output of the dynamic viscosity and the liquid fraction models is converted to effective viscosity using Eq. 2.

Each of the neural network models are trained using the Levenberg–Marquardt –algorithm. The training is stopped using the early–stopping method, in which the cross–validated error is minimized. The data itself is divided into training, validation and testing sets with the division of 70:20:10. However, it should be noted that as the neural networks are universal approximators, they are able to describe a noiseless system such as a data derived with computer experiments with practically negligible error and $R^2 \approx 0.99$.

Thus, the main variable that controls the network accuracy and its interpolation capabilities is the sampling density of the experimental design space, determined with the step size given in **Table 1**. Consequently, in the data generation, there forms a trade–off between the computational time of the thermodynamical, or other physical model –based software, and the accuracy of the surrogate–model. It should be noted that the extrapolation properties, ergo approximating values outside the design space with the surrogate–model, are not improved with a denser dataset (even with arbitrary small step size) within the determined compositional range.

The neural network describing the phosphorus partition is aggregated (ensemble aggregation) with multivariable linear regression models (a total of 7), in which the phosphorus partition ratio is described as a function of optical basicity and temperature. One of the models is based on a set of literature data ($n$ = 178), drawn from [10–18] and fitted against it. The model is given by

$$\log_{10} L_P = -44.0 + 57.1\Lambda + 61831 \cdot \frac{1}{T} - 79701.2\frac{\Lambda}{T} + 0.66 \ln\left[(\mathrm{FeO}) + 1\right], \quad (6)$$

where $\Lambda$ is the optical basicity and $T$ is the temperature in Kelvins. The rest of the aggregated models are reported in Refs. [14, 15]. The ensemble model prediction is simply the mean of the individual models. The ensemble models are known to be more stable than single models [21], and also in this case considered to give more realistic estimates of the system as the models probably provide information outside the FactSage database.

*2.5 The optimization algorithm*

The maximization of the objective function is carried out using a real–coded genetic algorithm (RCGA). The RCGA is programmed in–house, and it is implemented using the hybrid selection, recombination and mutation operators. The structure of the algorithm is essentially the same as in our previous work [22]. The operators are implemented using the following hybrid operators

1. Tournament selection & Roulette–wheel selection
2. Arithmetic & Convex crossover
3. Dynamic mutation & Mäkinen–Periaux–Toivanen mutation

In the case of hybrid operators, the probability of assigning an individual or a pair of individuals to either of the operators is P(operation) = 0.5. The crossover probability was selected as $P_{CX}$ = 0.9. The mutation probability evolved within the iterations according to the equation given in Ref. [23]. In addition, in each generation elitism operator was used to replace the individual with the lowest fitness level with the best individual. The convergence speed of the population towards the elite was also enhanced by applying the population mapping operator [24].

## 3. Results and discussion

*3.1 The optimization results*

In the scenario, the optimization variables were selected as CaO, $SiO_2$, FeO, $Al_2O_3$, MgO, conversion rate and input slag mass. The genetic algorithm population size was set to 50 and maximum number of generations to 200. The initial population and the constraints for the compositional variables were set as in **Table 1**. The initial phosphorus contents in the metal and slag phase were set 0.1 wt-% and 0 wt-%, respectively. In the initial state, the FeO content of the slag is set to 0 wt-%, and all the FeO in the dephosphorization state originates from the unreduced FeO. The reduced FeO ends up fully in the metal phase. The temperature was set to 1600 ∘C (1873.15 K). The constraints for optimization were

1. Slag/Metal -ratio $\leq 0.1$
2. Equilibrium phosphorus content in metal ≤ 20 ppm
3. Conversion rate ≥ 0.98
4. $\mu_{eff} \leq 1$ Pa·S

The liquid fraction of the slag was not constrained in the case. The optimization results are presented in **Table 2**. It can be seen that all the targets for optimization are satisfied. It should be noted that the repeatability of the results because of the stochastic nature of the RCGA was intentionally left on less notice. The slag composition that maximizes the phosphorus partition ratio within the given composition area and formulated constraints is the same as reported in Oeters [25], and thus the results are in good qualitative agreement with the literature. The $Al_2O_3$ and MgO contents in the slag were 0 wt-% in the optimal solution, implicating that the most favorable dephosphorization can be achieved in the ternary CaO–$SiO_2$–FeO system, and it is evident that a real-life slag cannot be considered free from these species. Nevertheless, in systematical experimental design, reduction of design variables simplifies the system and reduces the number of experiments needed. The need for design space simplification is emphasized in a novel process, such as HPSR with many potential covariates.

In the solution, the effective viscosity of the slag is well below the set constraint, the slag within this compositional region is not fully liquid. Thus, it might be that even though the slag might extract phosphorus efficiently in the thermodynamic point of view, the overall rate of reaction could be deteriorated because of the existing solid phase (≈23 wt-%). The slow rate of reaction might create a bottleneck in the full-scale process and demand the use of fluxes to achieve a fully liquid slag. Similar findings on the effect of time on the efficiency of the processing route has been reported in our previous work in the context of hot metal desulfurization [26]. However, the thermodynamic approach does not allow us to take the time component into account, and thus coupling of kinetic models within the optimization framework would be beneficial in the future endeavors.

**Table 2.** Optimization results for the formulated isothermal scenario.

| Reduction state | Value | Dephosphorization state | Value |
|---|---|---|---|
| Input FeO mass | 100000 kg | Output Fe mass [to metal] | 76416 kg |
| Input slag mass | 3977 kg | Output FeO mass (to slag) | 1691 kg |
| | | Conversion rate | 98.3% |
| | | Output slag mass | 5668 kg |
| | | Slag/metal -ratio | 0.07 |
| (CaO) | 77.1 wt-% | (CaO) | 54.2 wt-% |
| ($SiO_2$) | 22.9 wt-% | ($SiO_2$) | 16.0 wt-% |
| (FeO) | 0 wt-% | (FeO) | 29.8 wt-% |
| ($Al_2O_3$) | 0 wt-% | ($Al_2O_3$) | ≈0 wt-% |
| (MgO) | 0 wt-% | (MgO) | ≈0 wt-% |
| | | $B_2$ | 3.4 (-) |
| | | $\mu_{eff}$ | 0.05 Pa·s |
| | | $y_{liquid}$ | 0.77 |
| | | $L_P$ | 660 (-) |
| | | [P] | 20 ppm |

### *3.2 Interpretation of Results and Practical Implications*

**Figure 3** presents the effect of slag composition on the $L_P$, $\mu_{eff}$ and $y_{liquid}$ at 1600 °C (1873.15 K) The phosphorus partition ratio is maximized at the optimized composition, while the set constraint for the effective viscosity is satisfied. From the lower right corner, it can be observed that the fully liquid slag might compromise the thermodynamic dephosphorization potential of the slag, as the phosphorus partition ratio is only 1/6 (≈ 100) of the maximum observable value at the compositional region of the fully liquid slag. It should be noted that $Al_2O_3$ and MgO evidently end up in the slag phase because of dissolution from the refractory. However, according to computational results, the phosphorus partition ratio is maximized without these components in the slag phase.

Journal of Physics: Conference Series **3050** (2025) 012009 doi:10.1088/1742-6596/3050/1/012009

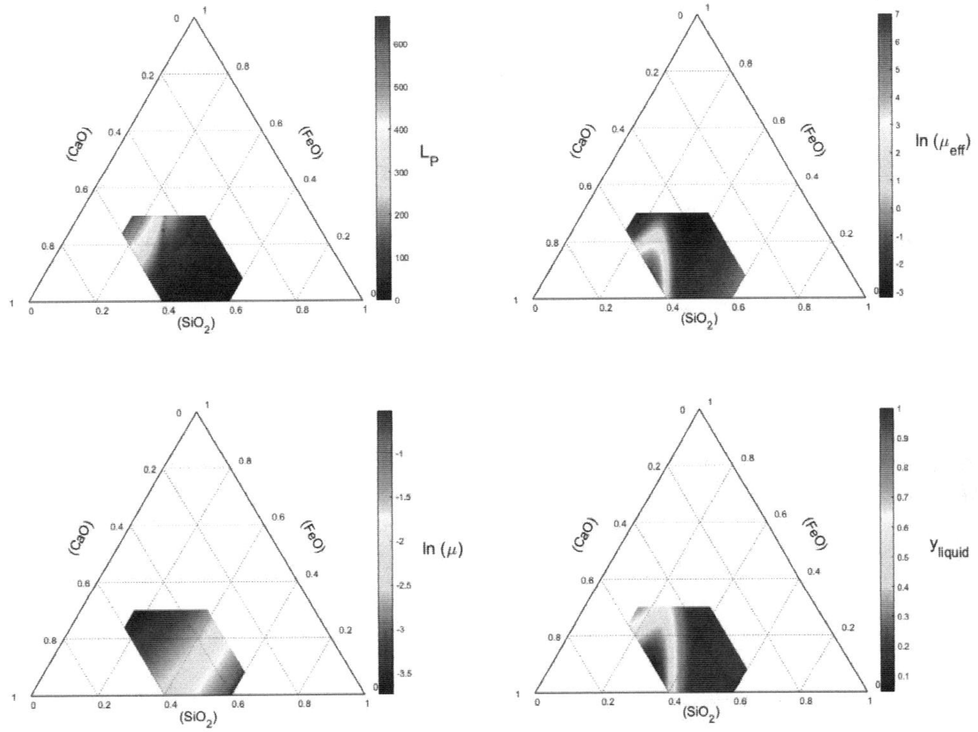


**Figure 3.** Ternary diagrams of the design space. The ternaries have been plotted at 1600 °C, $Al_2O_3$ = 0 wt-% and MgO = 0 wt-%.

The effect of temperature on the phosphorus partition ratio and effective viscosity was studied using the neural network models and the response surface method. The response surfaces are shown in **Figures 4a** and **4b**. From the figures it can be observed that the effect of temperature on the slag system is two-fold. On the other hand, increasing the temperature increases the liquid fraction of the slag, and consequently decreases the effective viscosity. Assumably the mass-transfer rate at the interface is also increased with the temperature. However, increasing the temperature drastically decreases the phosphorus partition ratio, and thus might have a negative net effect on the rate of dephosphorization as the thermodynamic driving force decreases. **Figure 4a** also illustrates the fact that the neural network that is trained based on FactSage data estimates the maximum $L_P$ to be achieved approximately at $B_2$ = 2.4. However, when computing the ensemble aggregated response, the $L_P$ maximum is obtained with $B_2$ = 3.45, which is consistent with the findings presented in Ref. [25]. Consequently, using the neural network instead of the ensemble model to predict the $L_P$ would drastically change the shape of the objective function, and thus the optimization result. In further applications, the output of the ensemble model could also be given as a weighted average, with the averaging parameters optimally determined, for example, using gradient boosting. Nevertheless, it should be noted that a careful machine learning model selection as well as the quantitative validation of the models with real-world data is fundamental while evaluating the computational results comprehensively. However, the independent quantitative validation of the approach demands extensive collection of data, which is out of the scope of this study.

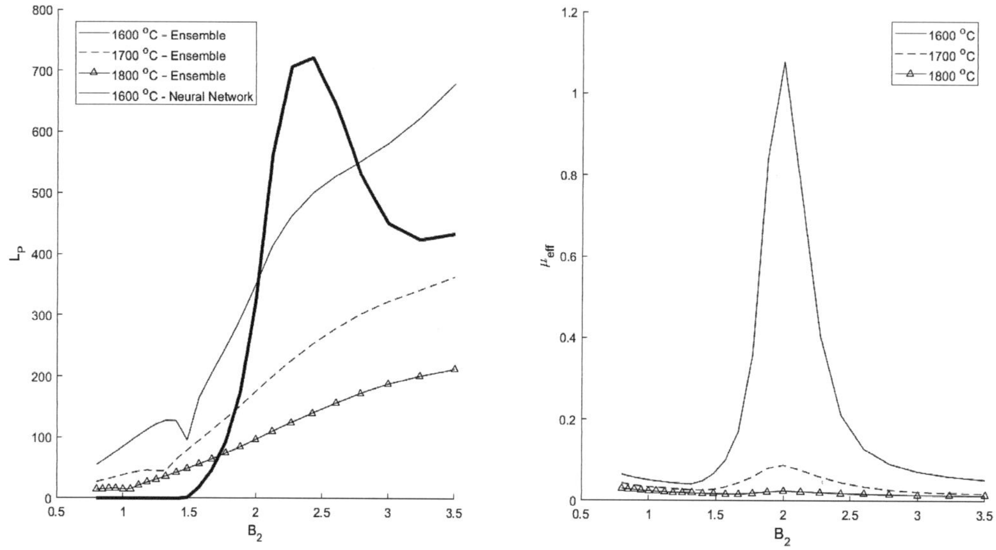


**Figure 4.** a) Effect of $B_2$ and temperature on the $L_P$ when FeO = 28 wt-%, and $Al_2O_3$ and MgO = 0 wt-%, b) Effect of $B_2$ on the $\mu_{eff}$.

**Figure 5** presents the effect of conversion rate and slag mass on the phosphorus partition ratio. With these assumptions, the conversion rate and the equilibrium phosphorus content form a pareto front, and the set phosphorus target in the metal phase constraints the achievable mass–balance –based conversion rate. This is because the FeO content in the slag phase improves drastically the dephosphorization capability of the slag along with the CaO content, and thus some iron oxides might be beneficial to be left unreduced to favor the phosphorus removal. It is also interpretable from the figure that the slag mass might constraint the dephosphorization. Because of practical aspects, the amount of slag cannot be increased infinitely, even though the slag mass is directly proportional to the mass of the extracted phosphorus.

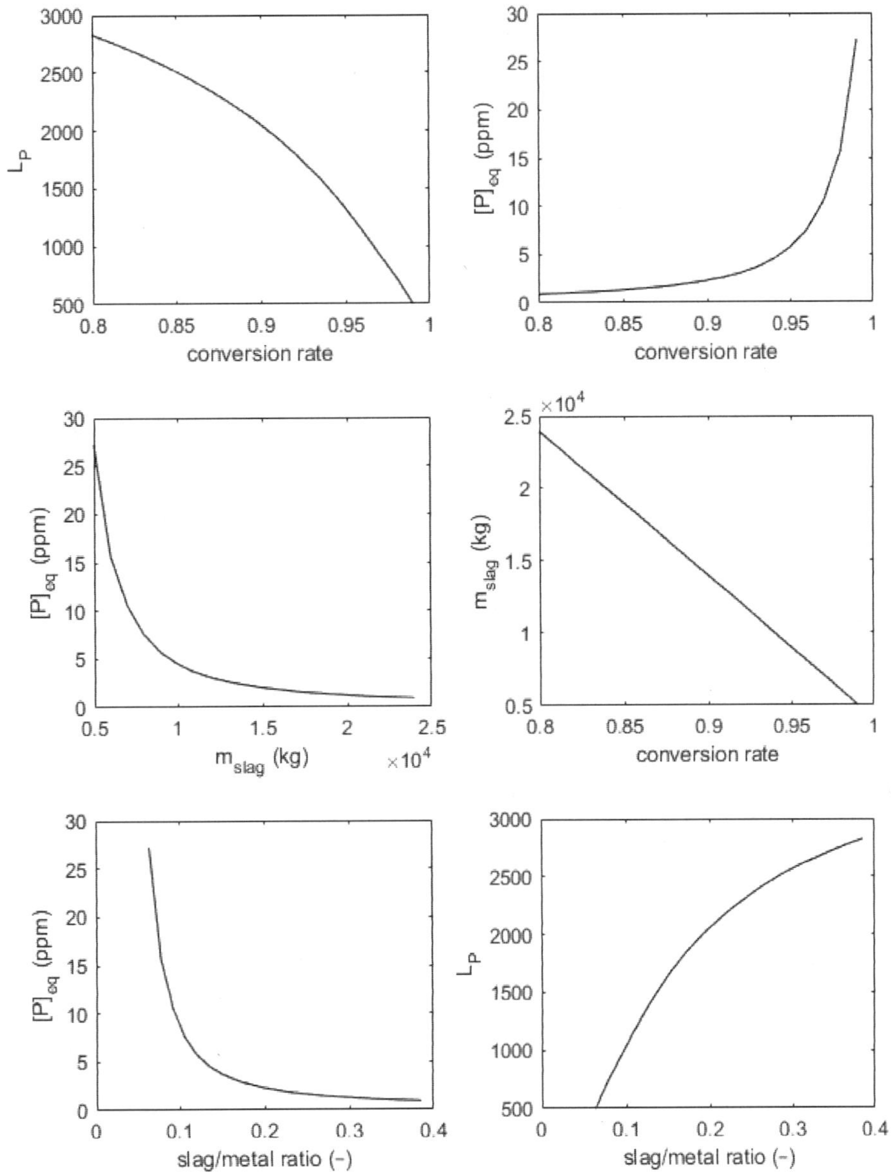


**Figure 5.** The effect of competing objectives (conversion rate) and constraints (slag/metal -ratio) on the dephosphorization behavior.

The modular structure algorithm allows the implementation of its components using any type of simulation model. However, using a sophisticated simulator to describe the complex system dynamics comes with increased computational cost. Nevertheless, the surrogate-model -based approach allows its use as a basis of model predictive control system, or similar. Coupling the optimization scheme for example with an advanced optical emission spectroscopy system would allow a real-time and predictive control of the slag composition in the furnace [27]. In addition, the surrogate-models can be integrated as a part of a dynamic model itself, as they provide a fast alternative for estimating thermodynamic properties and constraints for the time–dependent differential equations.

## 4. Conclusions

Hydrogen plasma smelting reduction is a novel process that can perform melting, reduction and refining in a single processing step. While the melting and reduction aspects have been studied at both laboratory and pilot scales, much less is known about the refining conditions, particularly regarding the removal of phosphorus.

To develop a tool for optimizing the dephosphorization parameters in the HPSR process, this study presents an optimization framework that enables rapid design of experiments and process design through a combination of computational thermodynamics and machine learning. It was found that the machine learning–based surrogate models provide a fast alternative for computing the objective function values, resulting in faster convergence of the optimization algorithm. The modular structure of the approach allows it to be coupled with dynamic reaction models and other types of surrogate models beyond shallow neural networks. However, more complex models naturally increase the computational complexity of the optimization. Nevertheless, a similar modeling and optimization framework can be applied to other design tasks as well.

The optimization results aligned well with the existing literature. It was found that a basicity of $B_2$ = 3.4 could be optimal for dephosphorization from a thermodynamic perspective, but the heterogeneous structure of the slag may reduce the dephosphorization rate. However, the stochastic nature of the RCGA might affect the repeatability of optimization. The optimization results are also inherently dependent on the selected models and the database used for training. Despite these limitations, the designed slag compositions can be used as a sophisticated initial guess for dynamic process experiments in future studies. Moving forward, research efforts would benefit from focusing on the kinetics of dephosphorization with heterogeneous slag structures to incorporate time dependency into the optimization framework.

## Acknowledgements

The research was supported by funding from the Sustainable Digitalized Metallurgy for a Climate-Neutral and Resource-Efficient Planet (SusMet4Planet) project, funded by the Austrian Research Promotion Agency; the Towards Carbon Neutral Metals 2 (TOCANEM 2) project, funded by Business Finland (no. 5372/31/2023); and the Hydrogen Plasma Reduction for Steelmaking and Circular Economy (H2PlasmaRed) project, funded by the European Union (no. 101138228).

## References

[1] Sabat, K. C., & Murphy, A. B. (2017). Hydrogen plasma processing of iron ore. *Metallurgical and Materials Transactions B*, *48*(3), 1561-1594.

[2] Souza Filho, I. R., Springer, H., Ma, Y., Mahajan, A., da Silva, C. C., Kulse, M., & Raabe, D. (2022). Green steel at its crossroads: Hybrid hydrogen-based reduction of iron ores. *Journal of Cleaner Production*, *340*, 130805.

[3] Sun, C., Wang, J., Zhou, M., Hong, L., Ai, L., & Wen, L. (2024). Process path for reducing carbon emissions from steel industry—Combined electrification and hydrogen reduction. *Processes*, *12*(1), 108.

[4] Naseri Seftejani, M., Schenk, J., Spreitzer, D., & Zarl, M. A. (2020). Slag formation during reduction of iron oxide using hydrogen plasma smelting reduction. *Materials*, *13*(4), 935.

[5] Sabat, K. C. (2022). Hematite reduction by hydrogen plasma: Where are we now?. *International Journal of Minerals, Metallurgy and Materials*, *29*(10), 1932-1945.

[6] Pauna, H., Ernst, D., Zarl, M., Aula, M., Schenk, J., Huttula, M., & Fabritius, T. (2022). Hydrogen plasma smelting reduction process monitoring with optical emission spectroscopy–Establishing the basis for the method. *Journal of Cleaner Production*, *372*, 133755.

[7] Büyükuslu, Ö. K., Aota, L. S., Raabe, D., Springer, H., & Souza Filho, I. R. (2024). Mechanisms and elemental partitioning during simultaneous dephosphorization and reduction of Fe-OP melts by hydrogen plasma. *Acta Materialia*, *277*, 120221.

[8] Naseri Seftejani, M., Schenk, J., Spreitzer, D., & Andreas Zarl, M. (2020). Slag formation during reduction of iron oxide using hydrogen plasma smelting reduction. *Materials*, *13*(4), 935.

[9] Thomas, D. G. (1965). A note on the viscosity of Newtonian suspensions of uniform spherical particles. *Journal of Colloid and Interface Science*, *20*, 267-277.

[10] Kovtun, O., Karbayev, M., Korobeinikov, I., Srishilan, C., Shukla, A. K., & Volkova, O. (2021). Phosphorus Partition Between Liquid Crude Steel and High-Basicity Basic Oxygen Furnace Slags. *Steel research international*, *92*(8), 2000607.

[11] Drain, P. B., Monaghan, B. J., Longbottom, R. J., Chapman, M. W., Zhang, G., & Chew, S. J. (2018). Phosphorus partition and phosphate capacity of basic oxygen steelmaking slags. *ISIJ International*, *58*(11), 1965-1971.

[12] Yang, X. M., Shi, C. B., Zhang, M., Duan, J. P., & Zhang, J. (2011). A thermodynamic model of phosphate capacity for CaO-SiO 2-MgO-FeO-Fe 2 O 3-MnO-Al 2 O 3-P 2 O 5 slags equilibrated with molten steel during a top–bottom combined blown converter steelmaking process based on the ion and molecule coexistence theory. *Metallurgical and Materials Transactions B*, *42*, 951-977.

[13] Lee, C. M., & Fruehan, R. J. (2005). Phosphorus equilibrium between hot metal and slag. *Ironmaking and steelmaking*, *32*(6), 503-508.

[14] Yang, W., Yang, J., Shi, Y., Yang, Z., Gao, F., Zhang, R., & Ye, G. (2021). Effect of basicity on dephosphorization of hot metal with a low basicity slag at 1653 K. *Ironmaking & Steelmaking*, *48*(1), 69-77.

[15] Huss, J., Berg, M., & Kojola, N. (2020). Experimental study on phosphorus partitions between liquid iron and liquid slags based on DRI. *Metallurgical and Materials Transactions B*, *51*(2), 786-794.

[16] Zhang, R., Jia, J., Li, S., Zhang, Y., Jing, L., Min, Y., & Liu, C. (2022). Analysis of Dephosphorization Thermodynamics Based on the Melt Structure of CaO–SiO2–FeO–MgO System. *ISIJ International*, *62*(8), 1586-1594.

[17] Ye, G. F., Yang, J., Zhang, R. H., Yang, W. K., & Sun, H. (2021). Behavior of phosphorus enrichment in dephosphorization slag at low temperature and low basicity. *International Journal of Minerals, Metallurgy and Materials*, *28*, 66-75.

[18] Van Niekerk, W. H., & Dippenaar, R. J. (1998). Phosphorus distribution between carbon-saturated iron at 1350 °C and lime-based slags containing $Na_2O$ and $CaF_2$. *Metallurgical and Materials Transactions B*, *29*, 147-153.

[19] Drain, P. B., Monaghan, B. J., Zhang, G., Longbottom, R. J., Chapman, M. W., & Chew, S. J. (2017). A review of phosphorus partition relations for use in basic oxygen steelmaking. *Ironmaking & Steelmaking*, *44*(10), 721-731.

[20] Yang, X. M., Li, J. Y., Chai, G. M., Duan, D. P., & Zhang, J. (2016). Critical evaluation of prediction models for phosphorus partition between CaO-based slags and iron-based melts during dephosphorization processes. *Metallurgical and Materials Transactions B*, *47*, 2302-2329.

[21] Breiman, L. (1996). Bagging predictors. *Machine learning*, 24, 123-140.

[22] Vuolio, T., Visuri, V.-V., Sorsa, A., Paananen, T., Tuomikoski, S., & Fabritius, T. (2023). Machine learning assisted identification of grey–box hot metal desulfurization model. *Materials and Manufacturing Processes*, *38*(15), 1983–1996.

[23] Bäck, T., & Schütz, M. (1996, June). Intelligent mutation rate control in canonical genetic algorithms. In *International Symposium on Methodologies for Intelligent Systems* (pp. 158-167). Berlin, Heidelberg: Springer Berlin Heidelberg.

[24] Arumugam, M. S., Rao, M. V. C., & Palaniappan, R. (2005). New hybrid genetic operators for real coded genetic algorithm to compute optimal control of a class of hybrid systems. *Applied Soft Computing*, *6*(1), 38-52.

[25] Oeters, F. (1994). Metallurgy of Steelmaking, Düsseldorf: Verlag Stahlheisen.

[26] Vuolio, T., Visuri, V.-V., Tähtilä, H., Pekuri, P., & Fabritius, T. (2024). The Synergistic Effect of Na2O on Hot Metal Desulfurization Kinetics in CaO-Na2O-SiO2-Al2O3-MgO Slag System. *Chemical Engineering Science*, *284*, 119525.

[27] Putaala, H. R., Pauna, H., Javed, A., Manzoor, U., Klapproth, D., Souza Filho, I. R., ... & Raabe, D. (2025). Effect of Furnace Parameters on Optical Emission Spectra of Hematite Reduction by Hydrogen Plasma. *Metallurgical and Materials Transactions B*, 1-19.

Journal of Physics: Conference Series **3050** (2025) 012010 doi:10.1088/1742-6596/3050/1/012010

# Model Predictive Control of the Refining Stage in an Electric Arc Furnace Using Soft Sensors

**Aljaž Blažič[1], Igor Škrjanc[1] and Vito Logar[1]**

[1]University of Ljubljana, Faculty of Electrical Engineering, Tržaška 25, 1000 Ljubljana, Slovenia

E-mail: aljaz.blazic@fe.uni-lj.si

**Abstract.**
In this paper, a novel Model Predictive Control (MPC) based on soft sensors is developed to optimize the refining stage in Electric Arc Furnaces (EAFs). Given the significant energy consumption and environmental impact associated with EAF operations, there is a critical need for more precise and efficient control strategies. The proposed MPC approach utilizes real-time soft sensors to accurately determine critical parameters, i.e., bath temperature and dissolved oxygen content. By replacing operators' subjective judgments with continuous, data-driven feedback, this methodology facilitates precise control adjustments throughout the refining process. The implementation of this MPC has shown that the duration of the refining stage has been reduced by approximately 80 seconds, which directly translates into shorter tap-to-tap times, lower energy consumption and improved operational efficiency. In addition, consistently maintaining optimal bath conditions ensures improved steel quality, greater reliability, and higher overall productivity of the EAF process.

## 1 Introduction

Steel is a vital material in modern industry, with no large-scale substitute available for its diverse applications. Currently, approximately 30% of global steel production comes from recycled steel, primarily processed using Electric Arc Furnaces (EAFs). While EAFs have less direct emission than blast furnace-basic oxygen furnaces (BF-BOF) process, their energy consumption and environmental footprint remain significant. Consequently, optimizing the EAF process has become a critical objective for the steel industry, with a focus on improving energy efficiency and reducing resource consumption [1].

In recent years, advancements in industrial digitalization have enabled significant progress in process modeling and optimization. Leveraging large volumes of operational data, combined with sophisticated modeling techniques, has created opportunities to enhance the performance of energy-intensive processes like EAF steel production. Comprehensive deterministic models for EAFs, based in fundamental physical principles, have been extensively developed to provide a detailed process description and facilitate the calculation of key performance variables. Notable theoretical modeling efforts include the work of Logar et al. [2, 3, 4, 5, 6], Fathi et al. [7, 8], Saboohi et al. [9], Meier et al. [10, 11], and Hay et al. [12, 13]. In parallel, data-driven methodologies have gained traction due to their capacity for leveraging operational data to enhance process performance. Machine learning techniques, such as linear regression, k-nearest neighbors (k-NN), and both evolving and conventional fuzzy methods, have been employed to predict and optimize electrical energy consumption in EAFs [14, 15]. Notably, optimization of electric arc performance using the Cassie–Mayr arc model has underscored its importance in enhancing arc efficiency and ensuring process sustainability [16]. Decision support systems incorporating real-time EAF optimizers have also been developed to improve furnace operation [17]. Specific applications of data-driven approaches include the use of fuzzy modeling to predict bath temperature [1] and dissolved oxygen content in the melt [18, 19]. Artificial neural networks (ANNs) have been applied to estimate slag foaming [20], molten steel temperature [21, 22], and carbon content [21, 22, 23].

Despite these advances, to the best of the authors' knowledge, no explicit control strategy for temperature or dissolved oxygen has been introduced in EAF operations. Controlling temperature during the refining stage remains a particular challenge as it has a direct impact on energy consumption, production efficiency, and steel quality. In addition, temperature measurements are significantly affected by a high level of noise and numerous outliers due to the non-homogenious bath temperature, making accurate control difficult. Control of dissolved oxygen content is also difficult, as accurate prediction of oxygen content is challenging and oxygen measurements themselves are subject to considerable noise.

Adaptive and predictive control are well established methodologies in the field of control systems engineering that address the limitations of traditional controllers by adapting to changes in the system or predicting future system behavior [24]. The model of a system can be used to make predictions about the system behavior, thereafter the model can be used to determine the optimal control actions that take the system dynamics and constraints into account, the approach known as predictive control [25]. Predictive controllers were originally designed for linear systems, but the idea has since been extended to nonlinear systems. In the following study [25], a two-degrees-of-freedom (2-DOF) model predictive controller incorporating a feed-forward component was proposed. The open-loop system model was identified using the evolving Fuzzy Model (eFuMo). Following the identification of the fuzzy model, it was integrated into the controller, which was subsequently activated. Upon activation of the 2-DOF controller, the identification process was deactivated, and the evolving fuzzy model was fixed to ensure stability. In [26], a fuzzy cloud-based predictive functional control (FCPFC) was developed and compared to the RECCo controller. The structure of the FCPFC was evolved in an online manner, while the consequent part was identified using the recursive weighted least squares (rWLS) algorithm. The results obtained from the Continuous Stirred Tank Reactor (CSTR) model demonstrated that the FCPFC controller exhibited a slightly faster response but a longer settling time compared to the RECCo controller. An adaptive Predictive Functional Controller (PFC) for hybrid continuous and discrete signals, based on the recursive least squares parameter identification method with exponential forgetting, is presented in [27]. This approach was examined on an exothermic batch reactor, a time-invariant multivariable process. The model predictive control (MPC) has shown a lot of promise in different fields, such as model-predictive trajectory-tracking control applied to a mobile robot in real time [28], helio-crain control [29], nonlinear plate heat exchanger [24], batch reactor that presents a complex hybrid–nonlinear–stiff systems [30] and tracking-error model-based predictive control for mobile robots [31]. The use of predictive models in process control offers a promising solution to address the challenges of controlling the refining stage in EAF. Accurate models can capture the nonlinear and dynamic behavior of complex systems, providing the foundation for advanced control strategies that improve process reliability and efficiency.

The EAF is a complex, nonlinear system that combines various electrical, chemical, mass-, and heat-transfer processes. To produce high-quality steel products, it is critical that the steel bath meets prescribed specifications prior to tapping. Achieving the desired steel grade requires precise control of bath temperature and carbon content. In the refining stage of the EAF process, operators currently measure the bath temperature and dissolved oxygen content only intermittently, using disposable probes. Each measurement necessitates switching off the furnace for approximately 25 seconds, causing operational delays, reduced energy efficiency, and productivity losses. Between these infrequent measurements, operators must rely on their experience and indirect indicators to estimate the current state of the bath. Since operator skill and experience vary widely, the frequency and accuracy of these estimates can differ significantly, resulting in inconsistencies and potentially unnecessary energy usage or bath overheating.

To mitigate these issues, this paper proposes Model Predictive Control (MPC) framework for bath temperature and dissolved oxygen control in EAFs. The MPC is based on the soft sensors that provide continuous and accurate estimations of bath temperature and dissolved oxygen content in real-time. By leveraging advanced modeling techniques and MPC, the soft sensors remove the operators from the critical decision-making loop, replacing subjective experience-based judgments with precise, data-driven estimates. This continuous monitoring allows for real-time adjustments to the EAF inputs, reducing the average time of refining stage by 80 seconds, which shortens tap-to-tap times, minimizes energy losses, and enhances operational efficiency. Employing MPC based on these soft sensor outputs ensures the optimal bath conditions, achieving the desired steel quality with improved reliability and consistency. Ultimately, integrating these predictive tools directly contributes to increased EAF productivity, improved steel quality, and greater overall plant efficiency.

## 2 Problem definition

EAF operators play a crucial role in achieving the desired steel quality during the steelmaking process. Precise control of bath melt during the refining stage is crucial to achieve the desired tapping temperature and carbon content. However, due to the chaotic and dynamic nature of the EAF process, it can be difficult to accurately determine the current state of the melt. Therefore, operators regularly measure the bath melt to monitor critical parameters such as temperature and dissolved oxygen content.

In many plants, these measurements require a temporary interruption of furnace operation, which can lengthen the production cycle and reduce the efficiency of the overall process. However, this is not always the case. Some modern EAF plants are equipped with advanced sensor systems, i.e., off-gas analysis and optical sensors mounted on water-cooled panel walls, which allow partial or continuous monitoring without interrupting the process. Nevertheless, accurate measurement of oxygen concentration in real time remains a challenge and is often only carried out towards the end of the refining process.

Inadequate control of melting parameters can lead to suboptimal steel quality and reduced mechanical performance. Controlling the temperature of the molten steel within the prescribed range prior to tapping is critical to achieving optimum steel quality and ensuring the suitability of steel properties for subsequent secondary metallurgy processes. Exceeding the required melting temperature range leads to excessive energy consumption, increases operational costs and further extends the process duration. In addition, too high oxygen content in the melt requires stronger deoxidation during secondary metallurgy, because if the oxygen is not properly removed, it contributes significantly to defects such as porosity and inclusions.

Consequently, this scenario presents a multi-input multi-output nonlinear control problem that requires advanced solutions to balance measurement accuracy, process efficiency, and product quality.

## 3 Model-based Predictive Control

Model-based Predictive Control (MPC) is an advanced control strategy that uses an explicit predictive model of the controlled process. The main advantage of MPC is the ability to predict future process states, which enables proactive rather than reactive control. MPC predicts the evolution of key control variables, such as bath temperature and dissolved oxygen content, over an extended future time horizon. These predictions, derived from soft sensor estimates, enable precise, data-driven control actions that effectively eliminate operator subjectivity and reliance on experience.

By continuously monitoring process conditions and dynamically adjusting inputs in real-time, MPC significantly optimizes the EAF operation. Such proactive adjustments help to shorten the duration of the refining stage, reduce overall tap-to-tap times, reduce energy consumption, and improve operational efficiency. The implementation of MPC based on soft sensor outputs ensures stable and optimal bath conditions, resulting in consistent steel quality with improved reliability and predictability.

### *3.1 Soft Sensors*

Soft sensors have been extensively studied and used for the accurate determination of bath temperature [1, 32] and dissolved oxygen content [18, 19] in EAF processes. Although comprehensive details on these sensors are beyond the scope of this article, a brief overview is provided below.

The bath temperature is estimated using a first-order discrete-time difference equation given by:

$$\hat{T}(k+1) = \hat{T}(k) + \frac{t_s}{m_{sc} C_{P,lSc}} \hat{P}(k) \tag{1}$$

where $\hat{T}(k)$ denotes the bath temperature at discrete time $k$, $t_s$ is the sampling time, $m_{sc}$ represents the scrap steel mass, $C_{P,lSc}$ is the specific heat capacity of the liquid steel, and $\hat{P}$ is the total applied power to the bath zone. The applied power $\hat{P}(k)$ is modeled using a Takagi-Sugeno fuzzy modeling approach as elaborated in [1]. During the refining stage, the total applied power to the bath zone consists of three contributing factors:

$$\hat{P}(k) = \hat{P}_{el}(k) + \hat{P}_{ch}(k) + \hat{P}_{ls}(k) \tag{2}$$

where $\hat{P}_{el}(t)$ represents electric arc power, $\hat{P}_{ch}(t)$ denotes the power generated by chemical reactions, and $\hat{P}_{ls}(t)$ accounts for thermal losses from the bath zone. The electric arc power $P_{el}$ depends primarily on transformer power $P_{tr}$ and the melt height, which correlates directly with the charged scrap mass $m_{sc}$. The chemical reaction power $P_{ch}$ arises from deviations from chemical equilibrium, which occur when carbon or oxygen is introduced into the EAF. Consequently, $P_{ch}$ directly depends on carbon mass flow $\dot{m}_C$ and oxygen mass flow $\dot{m}_{O_2}$. Lastly, the power loss $P_{ls}$ predominantly depends on the bath temperature $T(t)$ and the charged scrap mass $m_{sc}$.

The estimation of dissolved oxygen content follows a similar modeling approach described as:

$$\hat{\mathrm{O}}(k+1) = \hat{\mathrm{O}}(k) + t_s \sum_{i=1}^{M_{\mathrm{I}}} \Phi_{\mathrm{I},i}(\boldsymbol{u}_{\mathrm{O}}(k))\boldsymbol{\theta}_{\mathrm{I},i}^{\top}\boldsymbol{u}_{\mathrm{O}}(k), \tag{3}$$

where $\hat{\mathrm{O}}(k)$ denotes the dissolved oxygen content at time $k$, and the input vector $\boldsymbol{u}_{\mathrm{O}}(k)$ consists of the mass flows of carbon and oxygen, i.e., $\boldsymbol{u}_{\mathrm{O}}(k) = [\dot{m}_{\mathrm{O}_2}, \dot{m}_{\mathrm{C}}]^{\top}$. In this equation, $\boldsymbol{\theta}_{\mathrm{I},i}$ and $\Phi_{\mathrm{I},i}$ represent, respectively, the consequent parameters and validity function associated with the $i$th Instant Takagi-Sugeno model [19].

### *3.2 Structuring of the control law*

The structuring of the MPC includes the definition of the model-reference trajectory and formulation of the objective (criterion) function. Initially, the reference trajectory $\boldsymbol{w}$ must be established. For the scenario considered, this reference is defined by the last measurement prior to tapping: $\boldsymbol{w} = [T_m(N_m), \mathrm{O}_m(N_m)]^{\top}$, where $T_m$ represents the bath temperature measurement, $\mathrm{O}_m$ represents dissolved oxygen content, and $N_m$ denotes the total number of measurements taken during a heat. An interval reference, incorporating both lower and upper bounds, is then expressed as follows:

$$\underline{\boldsymbol{w}} = \boldsymbol{w} - \frac{1}{2}\Delta\boldsymbol{w}, \tag{4}$$

$$\overline{\boldsymbol{w}} = \boldsymbol{w} + \frac{1}{2}\Delta\boldsymbol{w}. \tag{5}$$

where $\Delta\boldsymbol{w}$ represents the allowable tolerance band for the reference trajectory.

For the present scenario, a first-order response trajectory characterized by a time constant $\tau_r$ serves as the model-reference trajectory. Over the defined prediction horizon, the deviation between the model-reference trajectory $\boldsymbol{r}(k)$ and the model-predicted output $\hat{\boldsymbol{y}}(k)$ is calculated as follows:

$$\boldsymbol{e}(k) = \boldsymbol{r}(k) - \hat{\boldsymbol{y}}(k). \tag{6}$$

where $\boldsymbol{r}(k) = [r_T(k), r_{\mathrm{O}}(k)]^{\top}$ and $\hat{\boldsymbol{y}}(k) = [\hat{T}(k), \hat{\mathrm{O}}(k)]^{\top}$ represent the model-reference trajectory and the predicted model outputs for the bath temperature and dissolved oxygen content, respectively.

The definition of the objective function for the optimization of the system inputs is of crucial importance and requires careful consideration. The control inputs are represented as $\boldsymbol{u}(k) = [P_{tr}, \dot{m}_{\mathrm{C}}, \dot{m}_{\mathrm{O}_2}]^{\top}$. The MPC aims to minimize the deviations from the desired model-reference trajectory while minimizing the variations of the control signal. The objective function for the MPC optimization problem is formulated as follows:

$$J = \sum_{k=N_1}^{N_2} \boldsymbol{e}(k)^{\top}\boldsymbol{Q}\boldsymbol{e}(k) + \sum_{k=0}^{N_u-1} \Delta\boldsymbol{u}(k)^{\top}\boldsymbol{R}\Delta\boldsymbol{u}(k), \tag{7}$$

where $\boldsymbol{Q}$ and $\boldsymbol{R}$ are constant weighting matrices that represent the relative importance of trajectory tracking errors and control input variations, respectively. The term $\Delta\boldsymbol{u}(k)$ represents the incremental change in control inputs between successive discrete time steps. The parameters $N_1$ and $N_2$ define the lower and upper boundaries of the prediction horizon, which form the coincidence horizon within which the output should exactly match the reference signal. The parameter $N_u$ specifies the number of future control actions to be optimized. The objective function is optimized using Particle Swarm Optimization (PSO) as detailed in the literature [33, 34, 35]. Following optimization, only the first element $\boldsymbol{u}(0)$ is applied to the system. At the subsequent discrete time instant, updated process output data become available, and the optimization procedure is repeated.

### *3.3 Simulation of measurements*

In MPC simulations, the entire refining stage is taken into account, with the first temperature and dissolved oxygen measurements serving as initial conditions. The simulation proceeds with the second measurement if both the estimated bath temperature and dissolved oxygen content fall within the predefined reference interval ($\underline{\boldsymbol{w}} \leq \hat{\boldsymbol{y}} \leq \overline{\boldsymbol{w}}$). Each simulated measurement includes a 25-second delay, whereby it is assumed that the measurement takes place in the middle of this delay.

It has been observed that the initial temperature measurements often contain numerous outliers, possibly due to bath inhomogeneity or incompletely melted scrap. Conversely, measurements taken later in the refining process show significantly fewer outliers, which is due to greater homogeneity of the bath and fully molten steel [1]. To improve reliability during the refining stage, an innovative backward simulation approach was introduced. In this approach, the soft sensor for the bath temperature is simulated in reverse chronological order, starting with the last measurement and working backwards to the first measurement. The temperature calculated by the soft sensor at the initial measurement point then defines the starting condition for the MPC. However, this calculated initial value is only determined after the next temperature measurement, so that the inherent noise of the initial measurements is effectively taken into account. A similar reverse-simulation method is also used to estimate the dissolved oxygen noise.

## 4 Results and discussion

### *4.1 Datasets*

Operational data were collected from an EAF located at a steelworks in Slovenia. The studied EAF has a capacity of 100 tons, is equipped with an 80 MVA AC transformer, and operates without auxiliary equipment such as gas burners, CO post-combustion systems, or stirring. The dataset used in this study includes measurements of total charged scrap mass ($m_{sc}$), transformer power ($P_{tr}$), carbon mass flow rate ($\dot{m}_C$), oxygen mass flow rate ($\dot{m}_{O_2}$), bath temperature ($T_m$), and dissolved oxygen content ($O_m$). The measurements of $m_{sc}$, $T_m$, and $O_m$ are event-based, while $P_{tr}$, $\dot{m}_C$, and $\dot{m}_{O_2}$ are sampled as continuous time series.

In total, data from 1900 heats were acquired. However, many dissolved oxygen measurements showed unrealistic values of 0 ppm due to measurement errors. These incorrect values were excluded along with excessively long heats due to maintenance work. After preprocessing, the dataset comprised 806 valid batches suitable for analysis. For the purposes of this study, the refining stage is defined as beginning with the first recorded bath temperature measurement.

### *4.2 Results*

Figure 1 presents a comparison of the control strategies used during the refining stage of the EAF. Subfigure 1 a shows the conventional approach based on operator-driven decision-making, while subfigure 1 b illustrates the proposed MPC strategy that uses soft sensors to estimate bath temperature and dissolved oxygen content. Although the MPC allows for flexible adjustment of inputs at each control step, the resulting profiles in subfigure 1 b appear relatively smooth and stable. This behavior reflects the optimization objective of MPC, which is to strike a balance between minimizing output error and reducing input variability while maintaining process constraints. The MPC dynamically adjusts control actions, enabling the desired refining conditions to be reached 184 seconds faster than with the conventional strategy, and with fewer measurements. The sudden jump observed at the second measurement point in subfigure 1 b is attributed to the noise in the initial measurement, as described in section 3.3.

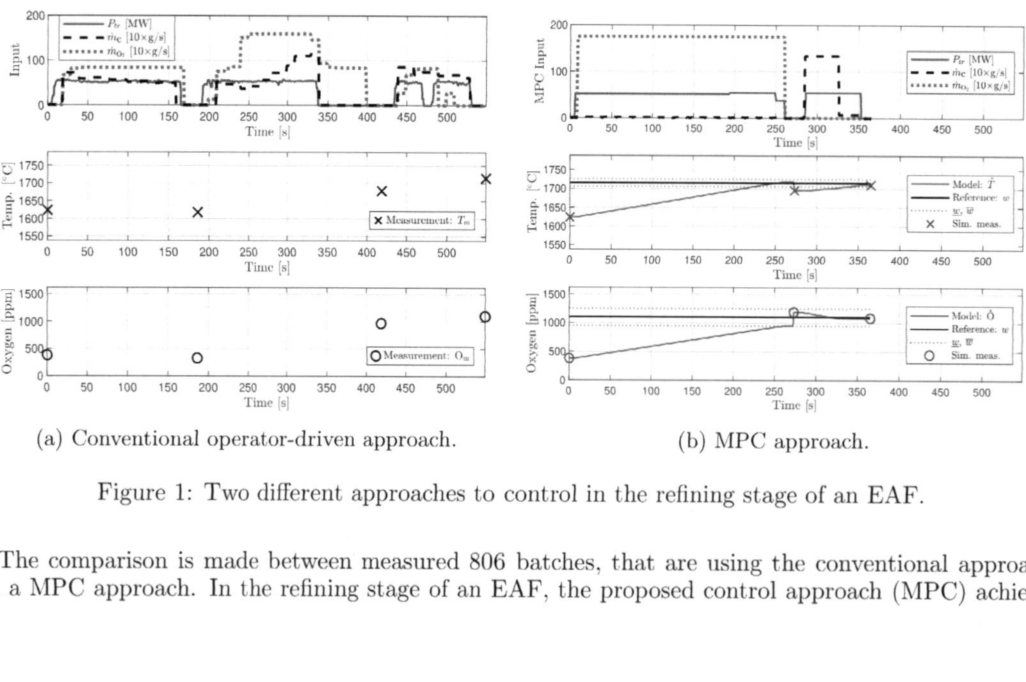


(a) Conventional operator-driven approach. (b) MPC approach.

Figure 1: Two different approaches to control in the refining stage of an EAF.

The comparison is made between measured 806 batches, that are using the conventional approach, and a MPC approach. In the refining stage of an EAF, the proposed control approach (MPC) achieves a significant improvement over the conventional approach. As shown in Table 1, the average batch time was reduced by 80 s, which corresponds to 15% of average relative time change.

| Metric | Value |
|---|---|
| Average Time Change | −80.3 s |
| Average Relative Time Change | −14.5 % |

Table 1: Comparison of conventional control and MPC in the refining stage of an EAF: average time and relative time changes.

### *4.3 Discussion*

In EAFs that operate with an AC transformer, the furnace power is controlled by discrete tap positions of the transformer. In industrial practice, the power is usually increased gradually by changing the taps in both the melting and refining stages. To ensure that the proposed MPC reflects realistic operational constraints, the tap position schedule has been limited so that it does not exceed that of the conventional control strategy. This constraint facilitates a fair and unbiased performance comparison by eliminating potential benefits resulting from an unrestricted tap selection in the MPC implementation. In addition, the carbon and oxygen mass flow were capped based on the maximum values observed in the conventional approach. However, further investigation is required to determine appropriate weighting matrices $\boldsymbol{Q}$ and $\boldsymbol{R}$ to effectively balance the power input, carbon and oxygen flow within the MPC approach.

Future work could integrate a Design of Experiment (DoE) methodology to strategically determine optimal timing for taking measurements, potentially enhancing model accuracy and system performance. Additionally, incorporating prediction intervals into the analysis may further refine the robustness and reliability of the control scheme.

## 5 Conclusion

This study presents a model predictive control system based on soft sensors to optimize the refining stage of electric arc furnaces. By providing real-time estimates of bath temperature and dissolved oxygen, the approach reduces the reliance on manual measurements and operator judgment. The proposed method reduces refining time by 80 seconds, increases energy efficiency, and improves the consistency of steel quality. These results demonstrate the effectiveness of data-driven control in increasing the productivity and process reliability of the EAF.

## Acknowledgment

The authors acknowledge the financial support from the Slovenian Research and Innovation Agency (Ph.D. grant for Aljaž Blažič, funding number P2-0219).

## References

[1] A. Blažič, I. Škrjanc, and V. Logar, "Soft sensor of bath temperature in an electric arc furnace based on a data-driven takagi–sugeno fuzzy model," *Applied Soft Computing*, vol. 113, p. 107949, 2021. [Online]. Available: https://doi.org/10.1016/j.asoc.2021.107949

[2] V. Logar, D. Dovžan, and I. Škrjanc, "Modeling and validation of an electric arc furnace: Part 1, heat and mass transfer," *ISIJ International*, vol. 52, pp. 402–412, 2012.

[3] ——, "Modeling and validation of an electric arc furnace: Part 2, thermo-chemistry," *ISIJ International*, vol. 52, pp. 413–423, 2012.

[4] V. Logar and I. Škrjanc, "Modeling and validation of the radiative heat transfer in an electric arc furnace," *ISIJ International*, vol. 52, pp. 1225–1232, 2012.

[5] ——, "Development of an electric arc furnace simulator considering thermal, chemical and electrical aspects," *ISIJ International*, vol. 52, pp. 1924–1926, 2012.

[6] V. Logar, A. Fathi, and I. Škrjanc, "A computational model for heat transfer coefficient estimation in electric arc furnace," *Steel Research International*, vol. 87, pp. 330–338, 2016.

[7] A. Fathi, Y. Saboohi, I. Škrjanc, and V. Logar, "Low computational-complexity model of eaf arc-heat distribution," *ISIJ International*, vol. 55, pp. 1353–1360, 2015.

[8] ——, "Comprehensive electric arc furnace model for simulation purposes and model-based control," *Steel Research International*, vol. 88, pp. 1–22, 2017.

[9] Y. Saboohi, A. Fathi, I. Škrjanc, and V. Logar, "Optimization of the electric arc furnace process," *IEEE Transactions on Industrial Electronics*, vol. 66, pp. 8030–8039, 10 2019. [Online]. Available: https://ieeexplore.ieee.org/document/8556389/

[10] T. Meier, K. Gandt, T. Echterhof, and H. Pfeifer, "Modeling and simulation of the off-gas in an electric arc furnace," *Metallurgical and Materials Transactions B: Process Metallurgy and Materials Processing Science*, vol. 48, pp. 3329–3344, 2017.

[11] T. Meier, A. H. Kolagar, T. Echterhof, and H. Pfeife, "Dynamic process modelling and simulation of an eaf and its dedusting system," *Chernye Metally*, vol. 137, pp. 18–24, 2018.

[12] T. Hay, T. Echterhof, and V. V. Visuri, "Development of an electric arc furnace simulator based on a comprehensive dynamic process model," *Processes*, vol. 7, pp. 1–17, 2019.

[13] T. Hay, V. V. Visuri, M. Aula, and T. Echterhof, "A review of mathematical process models for the electric arc furnace process," *Steel Research International*, vol. 92, pp. 1–22, 2021.

[14] S. Tomažič, G. Andonovski, I. Škrjanc, and V. Logar, "Data-driven modelling and optimization of energy consumption in eaf," *Metals*, vol. 12, 2022.

[15] G. Andonovski and S. Tomažic, "Comparison of data-based models for prediction and optimization of energy consumption in electric arc furnace (eaf)," *IFAC-PapersOnLine*, vol. 55, pp. 373–378, 2022.

[16] A. Blažič, I. Škrjanc, and V. Logar, "Arc quality index based on three-phase cassie–mayr electric arc model of electric arc furnace," *Metals*, vol. 14, 2024. [Online]. Available: https://www.mdpi.com/2075-4701/14/3/338

[17] S. Tomažič, I. Škrjanc, G. Andonovski, and V. Logar, "The development of simulation and optimisation tools with an intuitive user interface to improve the operation of electric arc furnaces," *Machines*, vol. 12, 2024.

[18] A. Blažič, V. Logar, and I. Škrjanc, "Fuzzy interval oxygen estimation in an electric arc furnace from scarce output measurements," in *IEEE International Conference on Fuzzy Systems (FUZZ-IEEE)*, 2022, pp. 1–8.

[19] A. Blažič, I. Škrjanc, and V. Logar, "Prediction interval soft sensor for dissolved oxygen content estimation in an electric arc furnace," *Applied Soft Computing*, vol. 167, p. 112246, 2024. [Online]. Available: https://doi.org/10.1016/j.asoc.2024.112246

[20] K. Son, J. Lee, H. Hwang, W. Jeon, H. Yang, I. Sohn, Y. Kim, and H. Um, "Slag foaming estimation in the electric arc furnace using machine learning based long short-term memory networks," *Journal of Materials Research and Technology*, vol. 12, pp. 555–568, 2021. [Online]. Available: https://doi.org/10.1016/j.jmrt.2021.02.085

[21] M. Niyayesh and Y. Uygun, "Predicting endpoint parameters of electric arc furnace–based steelmaking using artificial neural network," *International Journal of Advanced Manufacturing Technology*, 2024. [Online]. Available: https://doi.org/10.1007/s00170-024-14502-x

[22] C. Shi, L. Zhang, T. Zhou, Y. Wang, and Z. Ma, "Prediction of end-point temperature and carbon content in electric arc furnace steelmaking based on a whale optimization algorithm with lévy flight and stochastic configuration network algorithm," *Materials Today Communications*, vol. 42, p. 111326, 2025. [Online]. Available: https://doi.org/10.1016/j.mtcomm.2024.111326

[23] F. Yang, Z. Tang, Y. Shen, L. Su, and Z. Yang, "Parameter determination method of cassie-mayr hybrid arc model based on magnetohydrodynamics plasma theory," *Frontiers in Energy Research*, vol. 10, pp. 1–15, 2022.

[24] M. Ožbot and I. Škrjanc, *Evolving Fuzzy Model Predictive Control based on Optimization for Nonlinear Plate Heat Exchanger*, 2024, pp. 77–100.

[25] A. Zdešar, D. Dovžan, and I. Škrjanc, "A 2 dof predictive control based on evolving fuzzy model," *2014 IEEE Conference on Evolving and Adaptive Intelligent Systems, EAIS 2014 - Conference Proceedings*, pp. 1–6, 2014.

[26] G. Andonovski, E. Lughofer, and I. Škrjanc, "A comparison of recco and fcpfc controller on nonlinear chemical reactor," *Proceedings of the IASTED International Conference on Modelling, Identification and Control*, vol. 848, pp. 214–221, 2017.

[27] G. Karer, I. Škrjanc, and B. Zupančič, "Self-adaptive predictive functional control of the temperature in an exothermic batch reactor," *Chemical Engineering and Processing: Process Intensification*, vol. 47, pp. 2379–2385, 2008.

[28] G. Klančar and I. Škrjanc, "Tracking-error model-based predictive control for mobile robots in real time," *Robotics and Autonomous Systems*, vol. 55, pp. 460–469, 2007.

[29] A. Zdešar, O. Cerman, D. Dovžan, P. Hušek, and I. Škrjanc, "Fuzzy control of a helio-crane: Comparison of two control approaches," *Journal of Intelligent and Robotic Systems: Theory and Applications*, vol. 72, pp. 497–515, 2013.

[30] G. Karer, G. Mušič, I. Škrjanc, and B. Zupančič, "Hybrid fuzzy model-based predictive control of temperature in a batch reactor," *Computers and Chemical Engineering*, vol. 31, pp. 1552–1564, 2007.

[31] I. Škrjanc and G. Klančar, "A comparison of continuous and discrete tracking-error model-based predictive control for mobile robots," *Robotics and Autonomous Systems*, vol. 87, pp. 177–187, 2017. [Online]. Available: http://dx.doi.org/10.1016/j.robot.2016.09.016

[32] A. Blažič, I. Škrjanc, and V. Logar, "Estimation of the bath temperature in an electric arc furnace using operational data and fuzzy modelling approach," in *9th International Conference on Modeling and Simulation of Metallurgical Processes in Steelmaking (SteelSim 2021)*, 2021, pp. 1–8.

[33] J. Kennedy and R. Eberhart, "Particle swarm optimization," in *Proceedings of ICNN'95 - International Conference on Neural Networks*, vol. 4, 1995, pp. 1942–1948.

[34] E. Mezura-Montes and C. A. C. Coello, "Constraint-handling in nature-inspired numerical optimization: Past, present and future," *Swarm and Evolutionary Computation*, vol. 1, pp. 173–194, 2011. [Online]. Available: http://dx.doi.org/10.1016/j.swevo.2011.10.001

[35] M. E. H. Pedersen, "Good parameters for particle swarm optimization," *Hvass Lab., Copenhagen, Denmark, Tech. Rep. HL1001*, pp. 1551–3203, 2010.

Journal of Physics: Conference Series **3050** (2025) 012011 doi:10.1088/1742-6596/3050/1/012011

# Using a Novel Scaled Injector to Evaluate Biocarbon for Slag Foaming in EAF Steelmaking

**Christopher DiGiovanni[1*], Tiago Fernandes Lins[2], Michael Strelbisky[3], Majid Zamani[3], Ka Wing Ng[2], Xianai Huang[2], Allan Runstedtler[2], Colin Scott[1]**

[1] CanmetMATERIALS, Natural Resources Canada, 183 Longwood Road South, Hamilton, ON L8P 0A5, Canada
[2] CanmetENERGY, Natural Resources Canada, 1 Haanel Drive, Ottawa, ON K1A 1M1, Canada
[3] Tallman Technologies, 2220 Industrial St, Burlington, ON L7P 1A1, Canada

*E-mail: christopher.digiovanni@nrcan-rncan.gc.ca

**Abstract.** The steel industry faces increasing pressure to reduce carbon emissions, driving interest in sustainable alternatives to fossil-based carbon materials. Biocarbon, derived from renewable biomass, has shown promise as a charge carbon in electric steelmaking, but research into its application as an injection carbon for slag foaming is still ongoing. While studies have examined the reactivity and slag foaming behavior of biocarbon, its injectability has not been adequately evaluated due to the lack of a suitable experimental equipment. This study addresses this gap with a novel setup using a laboratory induction furnace and featuring a specially designed scaled-down industrial injector. Unlike previous experiments, which focused on evaluating slag foaming behaviour, this configuration enables the assessment of biocarbon injectability and slag foaming initiation under controlled laboratory conditions. CFD simulations of biocarbon injection helped inform both the injector design and the experimental conditions, ensuring effective penetration of the biocarbon particles. By closely replicating industrial injection conditions, this setup facilitates a new understanding of biocarbon's potential as a sustainable alternative in electric steelmaking.

## 1. Introduction

Iron and steelmaking is an extensive process that relies heavily on fossil fuels for refining of ores and to meet intense energy demands [1,2]. Thus, iron and steelmaking is one of the largest industrial emitters of $CO_2$, with global average emissions of ~2 tonnes of $CO_2$ per tonne of steel produced [3]. The global production of steel has grown dramatically over the last several decades which is expected to continue as steel market demand is tied to population and economic growth [4]. To meet the 2015 Paris Agreement climate change goals for 2050, steelmakers face the challenge of reducing $CO_2$ emissions and developing new technologies for sustainability.

In EAF steelmaking, it is common practice to inject carbon into the molten slag layer to generate a foamy slag, which increases the thermal efficiency of the process and protects the furnace refractory from the arc [5]. Additionally, slag foaming prevents reoxidation of the steel melt and reduces the absorption of nitrogen from the air. To create a foamy slag, the injection carbon acts as a reductant to the iron oxides (FeO) in the slag to produce CO gas bubbles. The CO gas can also reduce the FeO at the bubble interface to generate $CO_2$ [6]. Typically, injection carbon is petroleum coke or anthracite. However, to reduce $CO_2$ emissions from EAF steelmaking, biocarbon has received attention as an injection carbon material over the past several years.

Several studies have shown promise for using biocarbon as a foaming agent [7–11]. However, most have focused on the chemical reactions that generate CO bubbles to create foamy slag, leaving the injectability of biocarbon—especially compared to traditional fossil carbon—unexplored. A key challenge is whether biocarbon can be effectively injected and penetrate the slag layer. Because biocarbon is less dense than fossil carbon, it may not achieve sufficient momentum during injection to penetrate the slag layer and induce foaming. There are three possible outcome states for the biocarbon particles following injection. The first is penetration of the slag layer leading to the initiation of slag foaming, which would be considered successful injection. The other two are entrained in the off-gas flow or floating on the slag layer, neither of which would result in slag foaming and are deemed unsuccessful injection. These possible states are illustrated in Figure 1.

Biocarbon injection has been trialed in a few instances. For example, the GREENEAF2 project [12] trialed biocarbon injection in an industrial EAF and found that most of the biocarbon was entrained in the off-gas, resulting in poor foaming behavior. Echterhof et al. [13] conducted an industrial trial using palm kernel shells and observed no substantial change in foaming behavior based on visual inspection. However, injection in that trial was performed using a pneumatic door lance rather than a conventional carbon injector, which may have affected biocarbon penetration. Hoikkaniemi et al. [14] developed an experimental setup to study biocarbon as a slag foaming agent; but, it delivered biocarbon to the molten slag using a submerged quartz pipe, thereby limiting the study's ability to evaluate biocarbon injectability. To more specifically study biocarbon injectability, laboratory-scale tests specifically set up for injection experiments under controlled conditions are necessary. In this study, a novel scaled Tallman Supersonic Carbon

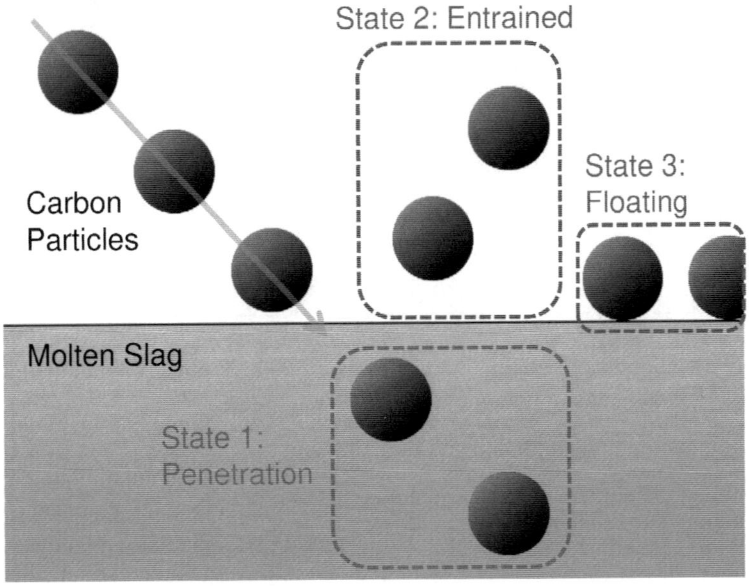


**Figure 1.** Illustration of the three possible outcome states for carbon injection: slag penetration, entrained in the off-gas, or floating on the slag surface.

Injector (TSCi) developed by Tallman Technologies, CanmetENERGY, and CanmetMATERIALS was employed. This injector uses a supersonic shrouding gas flow around the transport flow to enhance biocarbon penetration into the slag layer. Computational fluid dynamics (CFD) was used to model the particle flow from the injector, which guided the experimental setup. Observations during injection, measurements of slag height changes, and inspection of the solidified slag structure were used to determine the outcome state for biocarbon injection.

## 2. Materials and Methods

### *2.1 Materials*

#### *2.1.1 Injection Carbon*

A high quality biocarbon processed by pyrolysis from hardwood was used in this study and was provided by CanmetENERGY-Ottawa laboratory. The proximate and ultimate analysis was completed in accordance with the appropriate ASTM D7582, ASTM D5373, and ISO 562 standards and are shown in Table 1. The results show a relatively high fixed carbon (FC) content and relatively low ash content. The particle size of the injection biocarbon was controlled to a maximum of 325 μm and an average size of 145 μm.

**Table 1.** Chemical composition and properties for the biocarbon material studied

| Source | Proximate, db (wt%) | | | Ultimate, db (wt%) | | | | | Bulk Density ($kg/m^3$) |
|---|---|---|---|---|---|---|---|---|---|
| | FC | Ash | VM | C | H | N | S | O | |
| Hardwood | 83.10 | 1.48 | 15.41 | 86.60 | 2.67 | 0.48 | 0.05 | 8.78 | 325 |

#### *2.1.2 Slag Design*

A synthetic slag was prepared by melting a mixture of oxide powders in an appropriate ratio. The mixture was designed to match a nominal industrial slag composition. This target composition falls within the suitable range for slag foaming, based on its basicity and FeO and MgO contents [15]. Table 2 summarizes the oxide powders used, the composition of the powder mixture, and the target slag composition.

**Table 2.** Synthetic slag composition and $B_3$ basicity from mixing oxide powders

| Oxides | Target Slag (%) | Oxide Powder Mix (%) | Olivine Sand (%) | Lime Flux (%) | Iron Oxide (%) | Alumina (%) | Silica (%) |
|---|---|---|---|---|---|---|---|
| CaO | 37.00 | 39.48 | - | 98.50 | - | - | - |
| MgO | 9.62 | 10.50 | 42.00 | 1.00 | - | - | - |
| $SiO_2$ | 13.70 | 14.87 | 42.00 | 0.10 | - | - | 98.00 |
| $Al_2O_3$ | 6.00 | 6.41 | 0.70 | - | - | 96.00 | - |
| FeO | 24.53 | 26.22 | 9.00 | - | 98.00 | - | - |
| MnO | 5.50 | - | - | - | - | - | - |
| $P_2O_5$ | 0.56 | - | - | - | - | - | - |
| $TiO_2$ | 1.30 | - | - | - | - | - | - |
| Other | 1.79 | 2.52 | 6.30 | 0.40 | 2.00 | 4.00 | 2.00 |
| $B_3$ Basicity | 1.88 | 1.86 | - | - | - | - | - |

### *2.2 Experimental Setup & Method*

Melting was accomplished in a 100-kW induction furnace using an alumina crucible. A zirconia wash was applied to the surface of the crucible to reduce any interaction between the slag and the crucible. The dimensions of the alumina crucible were: ID of 140 mm and height of 250 mm. 8 kg of iron (<0.002 wt% C) was initially melted which acted as the heat source and stabilized the FeO present in the slag. Once the iron was molten (1625 $^0$C), 1 kg of slag was added to the melt. For the given alumina crucible ID, the charge resulted in 75 mm height of molten iron and 20 mm height of molten slag.

Once the slag was molten (~1600 $^0$C) and the temperature was stabilized, the carbon was injected into the molten bath. The scaled TSCi carbon injector has an annular flow shroud and a transport flow. The biocarbon powder met the transport air in the mixing chamber of the injector and flowed out through the nozzle along with the shroud air flow. A technical drawing of the scaled TSCi injector is shown in Figure 2.

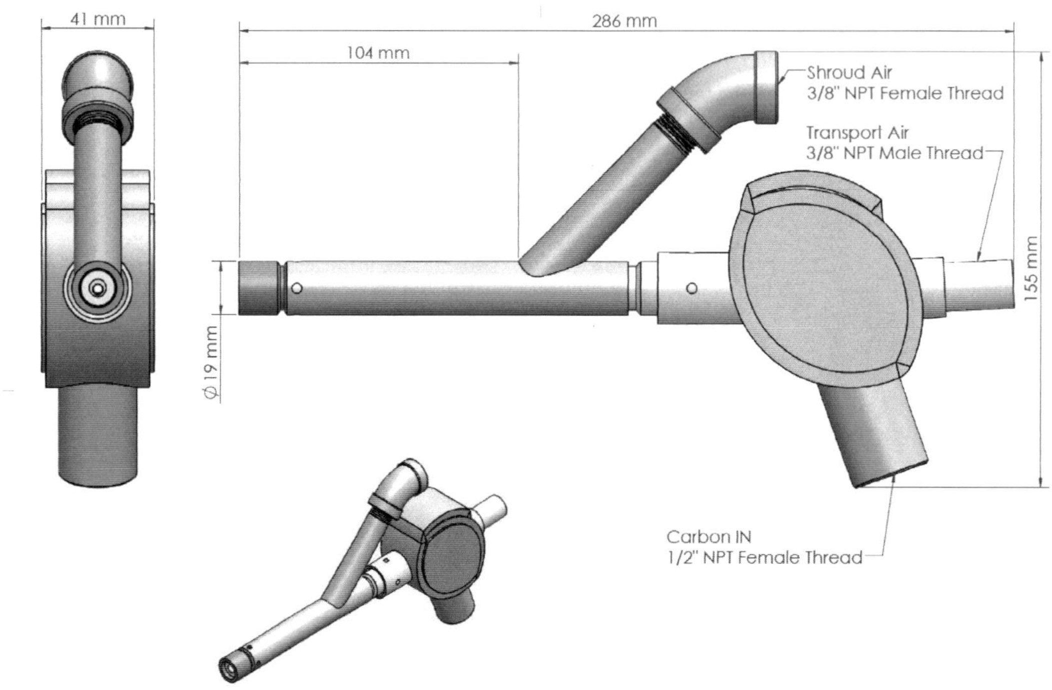


**Figure 2.** Solid model of the TSCi injector showing the exterior dimensions and isometric view.

The carbon injector was mounted on a rack and pinion system to easily position it over the centre of the crucible. The shroud and transport air flows were both set to 517 kPa using individual regulators at the injector. The experimental setup, including the crucible and injector, is shown in Figure 3(a) and a schematic of the crucible and melt is shown in Figure 3(b). Carbon particles were conveyed to the injector as a dilute phase using a powder feeder. The powder feeder was operating at 32 g / min and 10 LPM of carrier gas. The injector was positioned vertically over the crucible at a fixed distance from the slag surface and ran for 90 seconds. The melt was observed and monitored for signs of carbon penetration, entrainment, or floating during and after injection. Slag height was also physically measured after injection to determine if any slag foaming occurred

Journal of Physics: Conference Series **3050** (2025) 012011 doi:10.1088/1742-6596/3050/1/012011

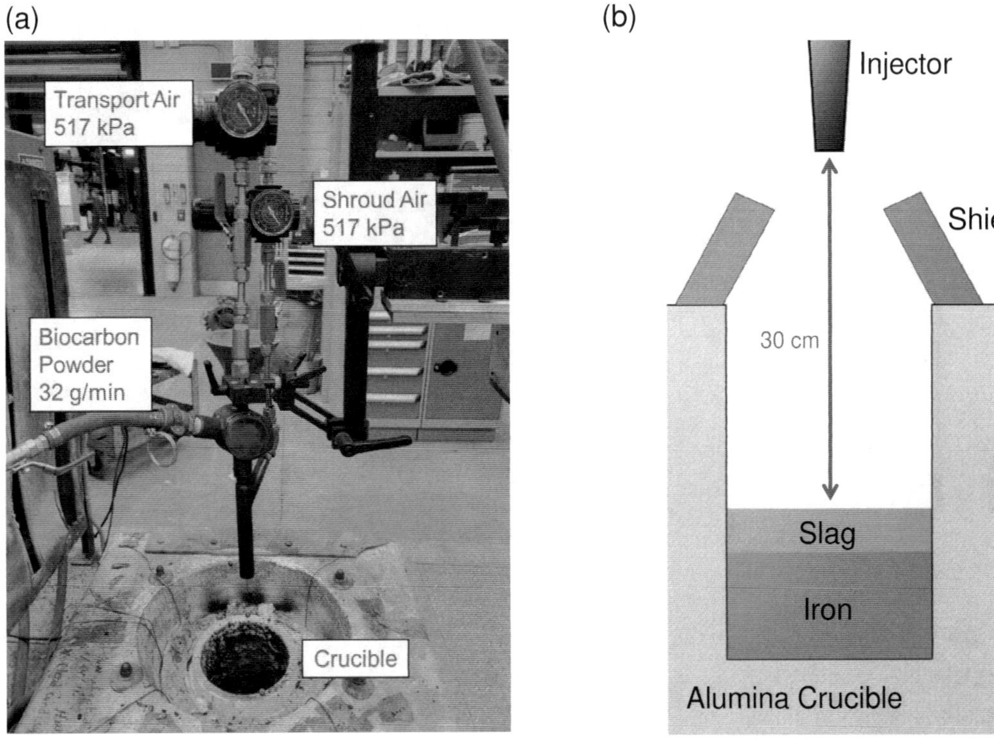


**Figure 3.** (a) Image of the experimental setup showing the scaled TSCi carbon injector and crucible (b) schematic of the crucible and molten material during testing.

### *2.3 Computational Fluid Dynamics*

A CFD model was developed to predict the particle and gas velocity profiles from the injection system. Simulations were conducted using Ansys Fluent 2023 R2. The simulation employed two inlet boundary conditions: injection carbon was supplied through the central nozzle with pressurized air at 517 kPa, and air was introduced through the shrouding nozzle at the same pressure. The domain outside the nozzle was set to atmospheric pressure.

The CFD domain was divided into three regions—the lance, the crucible, and the open zone between them (see Figure 4). The gas phase was modeled as a continuous phase composed solely of air. The fluid was treated as a compressible ideal gas, and turbulence was captured using the SST k-omega turbulence model. Equations for mass, momentum, and energy were solved in transient mode using a coupled pressure-based solver.

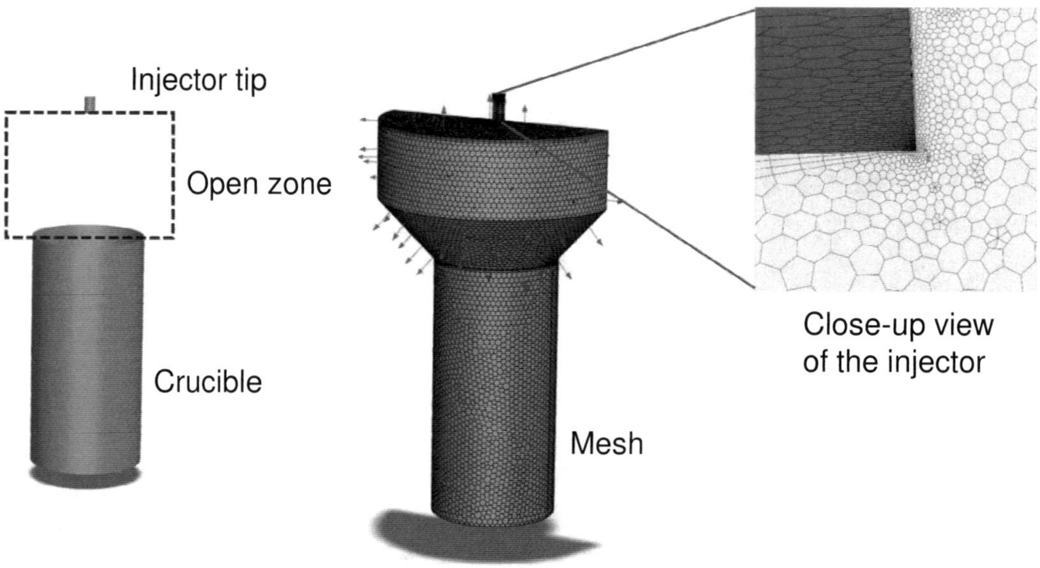


**Figure 4.** Design and schematic of the CFD model showing the general geometry and mesh density at the injector tip

Particle behavior was simulated using the Lagrangian particle tracking approach, with a modified spherical drag law to account for interactions between particles and Mach (1.84) flow of the transport air and shroud air. Two-way coupling between the particles and the gas phase was included in the model. Each particle was defined with a diameter of 150 μm and a density of 400 kg/m$^3$. The CFD geometry, as illustrated in Figure 4, comprised approximately 4 million cells.

## 3. Results and Discussion

### *3.1 CFD Analysis of Injector Performance*

To demonstrate the efficacy of the scaled TSCi injector to deliver carbon into the slag and the role of a shrouding jet, two cases were simulated: one with shrouding flow and one without, as shown in Figure 5. This was achieved by enabling or disabling flow into the annular boundary of the domain. The CFD results indicate that the shrouding jet significantly increased particle velocity, with particles traveling approximately 60 m/s faster than in the scenario without the shrouding jet as seen in Figure 5(b). In addition, Figure 5(c) shows the absence of a shrouding jet resulted in greater particle dispersion, an undesirable outcome when injecting carbon into the melt.

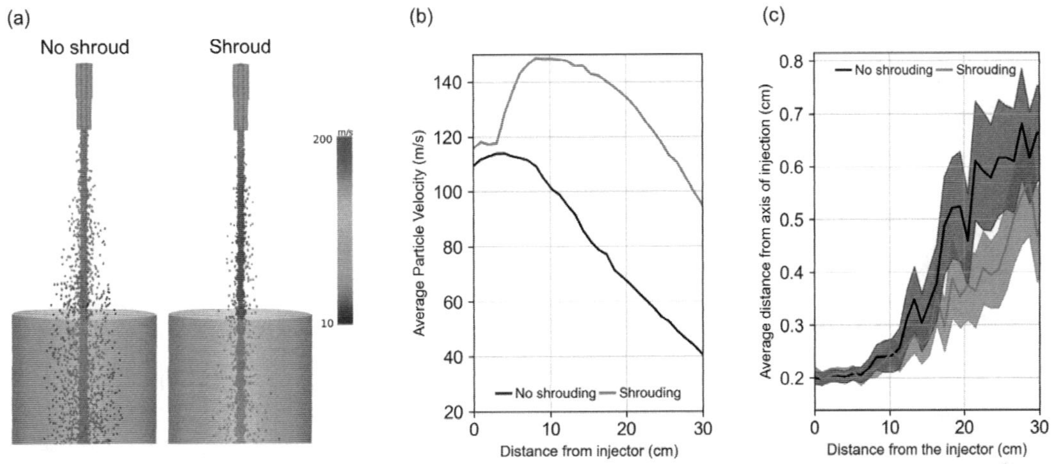


**Figure 5.** (a) Contour plot of the particle velocity with and without shrouding (b) average particle velocity and spread as particles travel from the injector tip showing the strong effect of shrouding.

The effect of the injector working distance (from injector tip to the slag surface) was also examined by modeling two cases with distances of 30 cm and 45 cm. The slag phase itself was not modeled; instead, its surface was defined as the bottom boundary of the domain. Figure 6(a) shows a contour plot of the particle velocity and Figure 6(b) shows the average particle velocities as they traveled from the injector toward the slag interface for both cases. The distance from the injector to the crucible does not appear to significantly impact the overall particle trajectory and speed; rather, a single curve describes the particle velocity versus distance from the injector, with the slag surface acting essentially as a cutoff. The results further demonstrate that the injector maintained particle velocities above 130 m/s at approximately 30 cm from the exit, which corresponds to its effective range. Beyond 45 cm, the average particle speed dropped below 75 m/s and radial dispersion increased significantly, which are believed to be detrimental to achieving slag penetration.

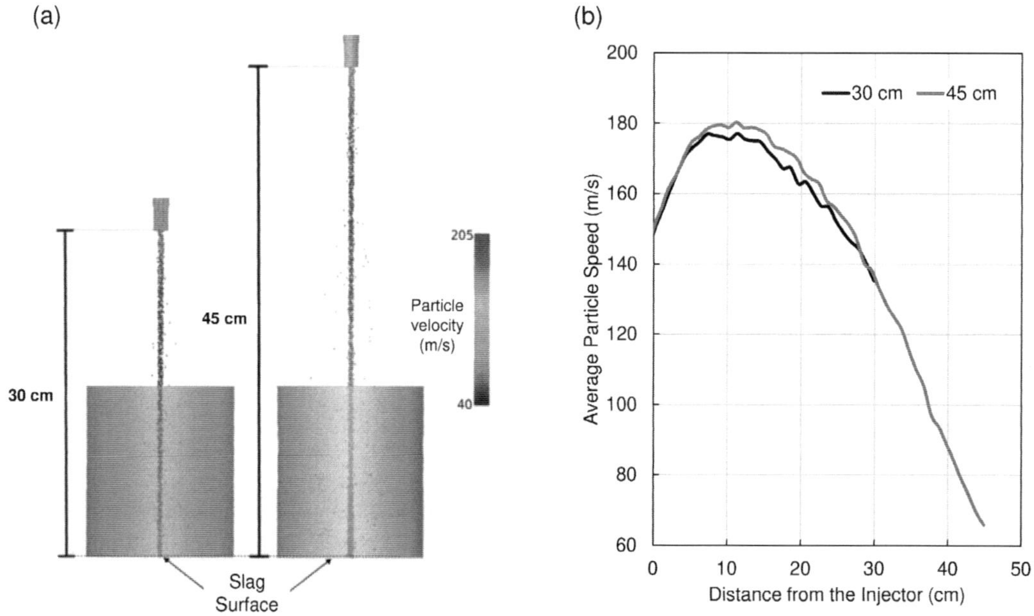


**Figure 6.** (a) Contour plot of particle velocity at working distances of 30 cm and 45 cm (b) average particle velocity at the two working distances of interest showing the steep drop in velocity.

Based on the CFD analysis, the experimental setup was optimized by carefully choosing the distance from the injector to the slag surface. The CFD data demonstrated that maintaining a distance of about 30 cm allowed the particles to retain velocities above 130 m/s, ensuring they had sufficient momentum to penetrate the slag. In contrast, beyond 45 cm the average particle speed dropped below 75 m/s and radial dispersion increased significantly, conditions that would likely reduce penetration efficiency and lead to entrainment or floating. The CFD simulations provided a clear, data-driven rationale for setting the initial distance at about 30 cm, ensuring that biocarbon particles retained the necessary momentum upon reaching the slag while minimizing dispersion. In this way, the CFD model enabled the fine-tuning of parameters and enhanced the reliability of the biocarbon injection trials.

### *3.2 Experimental Injection Performance and Slag Foaming*

Injection was carried out under the conditions determined by the CFD simulation. The injector tip was positioned vertically above the furnace at 30 cm from the slag surface. From the CFD results, this position should result in a particle velocity of approximately 130 m/s when reaching the slag surface. The experimental setup during injection is shown in Figure 7(a). During injection, minimal biocarbon was observed to be leaving the crucible or deviating from the axis of injection. This observation is a signal that the biocarbon particles were successfully contained within the air flow from the injector. Occasionally, a small amount of biocarbon particles would be ejected from the furnace with a minor amount of slag. It is postulated this is related to aggressive reactions and bubbling during injection. Figure 7(b) shows the slag surface directly after injection. From the figure, it can be observed that there is no biocarbon floating on the slag surface, which would be visible as mounds of black dust. Tests at lower particle speed (both lower pressure and larger working distance) showed a significant amount of biocarbon on the slag surface. During those experiments, biocarbon floating on the slag surface was also consistently ejected from the crucible by the air flow generated by the injector. By using optimized injection parameters in the presented experiment, no biocarbon particles were observed on the slag surface and a minimal amount of biocarbon was observed leaving the furnace.

(a) (b)

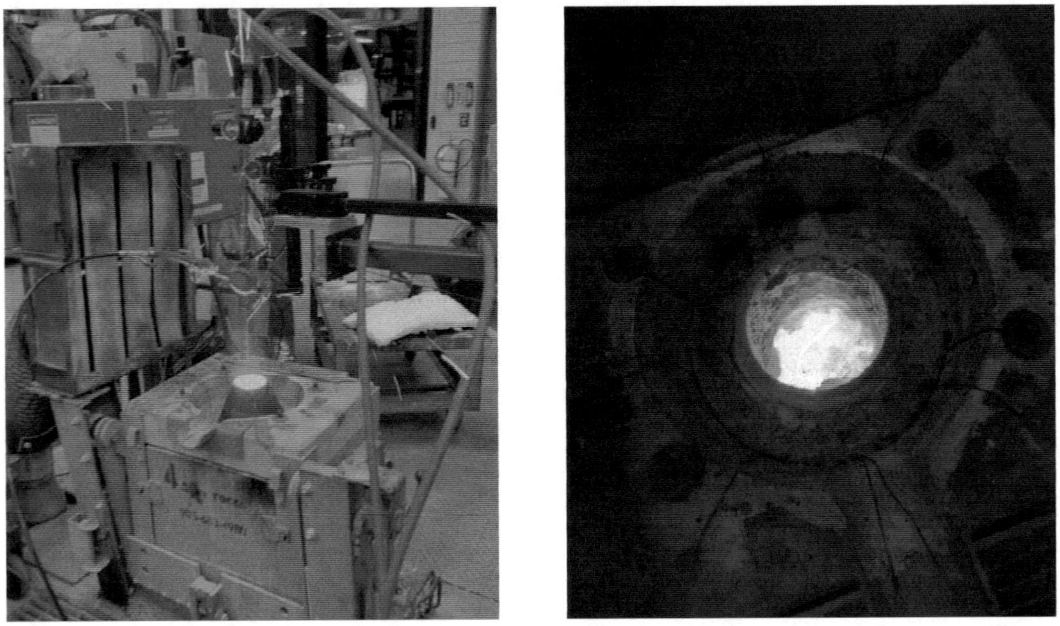

**Figure 7.** (a) Photo of the experimental setup while biocarbon was being injected into the crucible (b) image of the slag surface directly after injection showing no signs of biocarbon floating on the surface.

Following injection, slag was observed along the interior walls of the crucible, indicating significant movement of slag during injection, likely due to the injector air stream or carbon reactions. The peak height of the slag surface was measured 2 minutes after injection stopped and had increased by 182% compared to the original height. A second measurement, taken approximately 10 minutes after injection, showed a 111% increase from the original height. It should be noted that power to the furnace was shut off prematurely at 4 minutes after injection, which likely discouraged more intense slag foaming. Nonetheless, the change in height indicates that slag foaming initiated during injection and continued after injection was stopped. The biocarbon particles were able to penetrate the slag layer, react with the FeO in the slag, and cause slag foaming.

After the melt solidified and cooled, the slag was inspected for signs of slag foaming and biocarbon penetration. Slag foaming occurs when CO bubbles become trapped in the slag, leading to volume expansion. These bubbles are the direct result of FeO reduction by the injected carbon. Fine bubbles (~1 mm) are considered effective in creating and sustaining a foamy slag [16,17]. Since these bubbles form only when the carbon particles penetrate the slag and react with FeO, their presence in the solidified slag serves as clear evidence of carbon penetration. Figure 8(a) shows an overall view of the solidified materials from the crucible experiment, while Figures 8(b) through 8(d) offer detailed views of the bubbles. The bubbles range from 0.5 mm to 5 mm in diameter, with a common size of ~1 mm, which confirms the reaction occurred and demonstrates biocarbon penetration during injection.

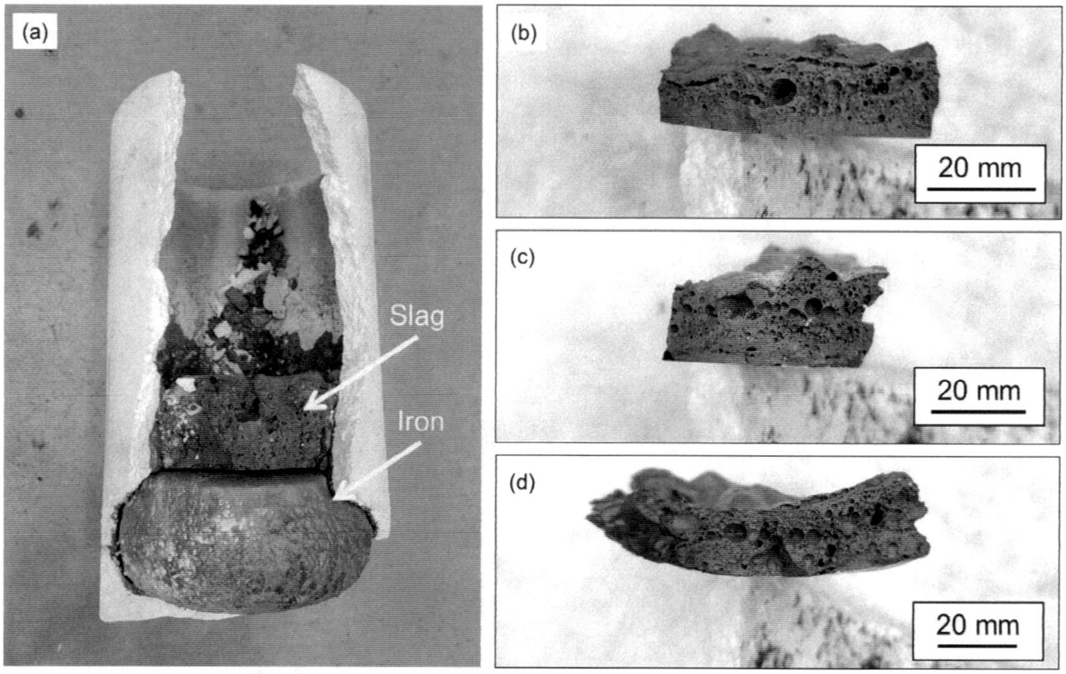


**Figure 8.** (a) Image of the solidified iron and slag showing many small bubbles within the slag structure (b) – (d) detailed views of the bubbles within the solidified slag.

The measured increase in slag height and the observed fine bubbles in the slag provide clear evidence of biocarbon particle penetration. The injected biocarbon particles penetrated the slag layer and reacted with the FeO, producing CO bubbles that resulted in slag foaming. Therefore, the biocarbon injection outcome state in these experiments can largely be characterized as penetration of the slag. Despite the relatively low density of biocarbon, the specialized scaled TSCi injector provided sufficient momentum to the particles to penetrate the slag layer and avoid entrainment or floating.

These experiments represent a significant advance in laboratory-scale investigations into biocarbon injectability. Previous studies either added biocarbon in packets, stirring them into the molten slag, or used a submerged lance [18]. While these methods allowed for examination of the chemical effects of biocarbon on slag foaming, the injection process itself was assumed rather than directly tested. Being able to inject biocarbon into the melt enables direct comparison of the injection behavior of various carbon types, a critical factor for its industrial adoption. Rather than immediately resorting to large-scale industrial trials, promising carbon candidates can be studied and qualified at the laboratory scale, thereby streamlining the development process and reducing risks for future applications.

## 4. Conclusions

This study demonstrated the potential of biocarbon injection as an effective foaming agent in EAF steelmaking using a novel scaled injector. CFD analysis revealed that the shrouding jet significantly boosted particle velocity and minimized dispersion, ensuring that biocarbon particles achieved sufficient momentum to penetrate the slag layer. Experimental observations confirmed these findings, with no biocarbon accumulation on the slag surface and minimal ejection from the crucible. The marked increase in slag height and the presence of fine gas bubbles provided clear evidence of successful CO generation through FeO reduction.

These results validated the laboratory-scale feasibility of biocarbon injection, offering a reliable framework to assess alternative carbon sources prior to full-scale industrial trials. Future work should focus on further optimizing the experimental procedure and biocarbon properties to enhance process efficiency and support the sustainable decarbonization of the steelmaking industry.

## Acknowledgements

Experiments were conducted at CanmetMATERIALS with the valued contributions of Bruce Ingram and Peter Newcombe. The authors also gratefully acknowledge Nicole Perna and Graeme Scott of ArcelorMittal Dofasco for providing their industrial expertise and materials. This work was supported by Office of Energy Research and Development Canada.

## References

[1] International Energy Agency 2015 *Coal Information* (Paris, France: IEA Publications)

[2] International Energy Agency 2015 *Energy Statistics of Non-OECD Countries 2015* (Paris, France: IEA Publications)

[3] Daehn K, Basuhi R, Gregory J, Berlinger M, Somjit V and Olivetti E A 2022 Innovations to decarbonize materials industries *Nat. Rev. Mater.* **7** 275–94

[4] Raabe D 2023 The Materials Science behind Sustainable Metals and Alloys *Chem. Rev.* **123** 2436–608

[5] Morales R D, Romero J A, Rubén L G, LÓpez F and Camacho J 1995 The Slag Foaming Practice in EAF and Its Influence on the Steelmaking Shop Productivity *ISIJ Int.* **35** 1054–62

[6] Sarma B, Cramb A W and Fruehan R J 1996 Reduction of FeO in Smelting Slags by Solid Carbon: Experimental Results *Metall. Mater. Trans. B Process Metall. Mater. Process. Sci.* **27** 717–30

[7] Wei R, Zheng X, Zhu Y, Feng S, Long H and Xu C C 2024 Hydrothermal bio-char as a foaming agent for electric arc furnace steelmaking: Performance and mechanism *Appl. Energy* **353** 122084

[8] Yunos N F M, Ismail A N, Munusamy S R R, Idris M A and Sahajwalla V 2021 Reaction Kinetics of Palm Char and Coke with Iron Oxides in EAF Steelmaking Slag *J. Sustain. Metall.* **7** 412–26

[9] Fidalgo B, Berrueco C and Millan M 2015 Chars from agricultural wastes as greener fuels for electric arc furnaces *J. Anal. Appl. Pyrolysis* **113** 274–80

[10] Kieush L, Schenk J, Koveria A, Rantitsch G, Hrubiak A and Hopfinger H 2023 Utilization of Renewable Carbon in Electric Arc Furnace-Based Steel Production: Comparative Evaluation of Properties of Conventional and Non-Conventional Carbon-Bearing Sources *Metals (Basel).* **13**

[11] Han C, Wei G, Zhu R, Cheng Y, Wang R, Wu W and Wang Y 2023 Characterization and kinetics of iron oxide reduction by corn straw biochar prepared by pyrolysis using superheated steam atmosphere and hydrothermal processes *Renew. Energy* **219** 119571

[12] Baracchini G, Bianco L, Cirilli F, Echterhof T, Griessacher T, Marcos M, Mirabile D, Reichel T, Rekersdrees T and Sommerauer H 2018 *Biochar for a sustainable EAF steel production (GREENEAF2)* (Brussels, Belgium)

[13] Echterhof T, Demus T, Pfeifer H, Schlinge L and Schliephake H 2016 Investigation of palm kernel shells as a substitute for fossil carbons in a 140 t DC Electric Arc Furnace *Proc. 11th Eur. Electr. Steelmak. Conf. \& Expo. 11th Eur. Electr. Steelmak. Conf. \& Expo, Venice, Italy* 25–7

[14] Hoikkaniemi E, Sulasalmi P, Visuri V-V and Fabritius T 2024 Biochar as a slag foaming agent in EAF – A novel experimental setup *IOP Conf. Ser. Mater. Sci. Eng.* **1309** 012010

[15] Pretorius E B and Carlisle R C 1998 Foamy Slag Fundamentals and Their Practical Application to Electric Furnace Steelmaking *Proceedings of the 16th Process Technology Conference* (New Orleans, LA, USA)

[16] Barella S, Gruttadauria A, Mapelli C and Mombelli D 2012 Critical evaluation of role of viscosity and gas flowrate on slag foaming *Ironmak. Steelmak.* **39** 463–9

[17] Zhang Y and Fruehan R J 1995 Effect of the bubble size and chemical reactions on slag foaming *Metall. Mater. Trans. B* **26** 803–12

[18] DiGiovanni C and Echterhof T 2024 Progress Toward Biocarbon Utilization in Electric Arc Furnace Steelmaking: Current Status and Future Prospects *J. Sustain. Metall.*

# Numerical Investigation of Hydrogen Blending on the Impinging Flame Structure in Non-Premixed $CH_4/H_2/Air$ combustion for Scrap Metal Heating

Gopal Pandey[1,2,*], Geoffrey Brooks[1], Jamal Naser[1], Daniel Liang[2]

[1]Victorian Hydrogen Hub (VH2), Swinburne University of Technology, Hawthorn, VIC, Australia

[2]Commonwealth Scientific and Industrial Research Organisation (CSIRO), Clayton, VIC, Australia

[*]Corresponding Author: Gopal Pandey, gopalpandey@swin.edu.au

## Abstract:

Gas burners play a crucial role in various ironmaking and steelmaking processes, particularly for heating and cutting operations. In Electric Arc Furnaces (EAFs), high-speed gas burners are widely used to enhance thermal efficiency. While the majority of heat in EAF is generated by electric arcs, gas burners help distribute heat more uniformly, improving overall energy efficiency. Currently, most of these burners operate with natural gas (primarily methane ($CH_4$)) as fuel and oxygen or air as oxidiser. The gases are supplied through separate ports, forming non-premixed flames. As these flames impinge on scrap metal inside the furnace, heat is transferred primarily through convection and radiation to the scrap metal. However, despite their efficiency, these flames contribute to carbon dioxide ($CO_2$) emissions, increasing the furnace's overall carbon footprint. With the steel industry striving to reduce carbon emissions, hydrogen is emerging as a promising alternative fuel-particularly green hydrogen, which is produced with zero carbon emissions. Due to challenges associated with hydrogen transportation and storage, blending hydrogen with natural gas is becoming an economical transition strategy in the early stages of decarbonisation. However, before implementing this approach, it is essential to understand the combustion and heat transfer characteristics of hydrogen-enriched non-premixed impinging flames. Despite its significance, research in this area remains limited. This study aims to investigate the combustion and heat transfer behaviour of non-premixed flames impinging on a steel plate. A numerical approach using computational fluid dynamics (CFD) modelling, coupled with a thermodynamic combustion mechanism, is employed to analyse the flow field, combustion zone, and heat transfer characteristics. Large Eddy Simulation (LES) with the GRIMech3.0 combustion mechanism is applied to study different cases with 10 pct, 25 pct, 50 pct, and 75 pct hydrogen blending with natural gas and hydrogen flames. In all cases, the burner geometry remained identical, with only flow variables altered. The findings indicate that, despite hydrogen-blended fuels being supplied at a higher velocity than that of methane to maintain equivalent heat release, it results in approximately the same wall jet region as methane-impinging flame. The results show that the temperature of the flame increases as the hydrogen content increases in the flame. Also, flames with 100 pct hydrogen fuel are predicted to heat the steel plate to $735K$ compared to $665K$ for methane flame.

Keywords: **Electric Arc Furnace, Scrap Heating, Hydrogen Combustion, CFD, Impinging Jet Flames**

## 1. Introduction

In Electric Arc Furnaces (EAFs), gas burners play a vital role in providing extra heat to cold spots, improving the furnace's overall thermal efficiency [1-5]. Gas burners are known to enhance productivity by 5 to 20 pct [4, 6]. These gas burners typically direct flames onto piles of scrap within the furnace, creating an impinging flame effect. The way heat transfers in an impinging jet directly influences the overall heat exchange between the flame and the furnace[2, 6]. Additionally, these burners have distinct features—supersonic oxygen injection, for example, is essential in steelmaking for metallurgical processes[7]. To optimise performance, high-speed burners are often combined with oxygen injection systems to create hybrid burners that can switch between different operating modes, such as flame-generating mode and coherent oxygen injection mode[8]. Because of these functionalities, fuel and oxidiser are supplied separately, classifying the burner as a non-premixed system. This design not only enhances performance but also improves safety by reducing the risk of flashback.

When a flame impinges on a surface, it forms an impinging jet flame, with heat transfer characteristics governed by both the flame properties and the fluid dynamics along the surface. The flow structure of an impinging jet flame is typically divided into three regions: the free jet region, the stagnation region, and the wall jet region. The stagnation region and wall jet region are collectively referred to as the impingement zone. Figure 1 provides a schematic representation of these flow regions in an impinging jet flame.

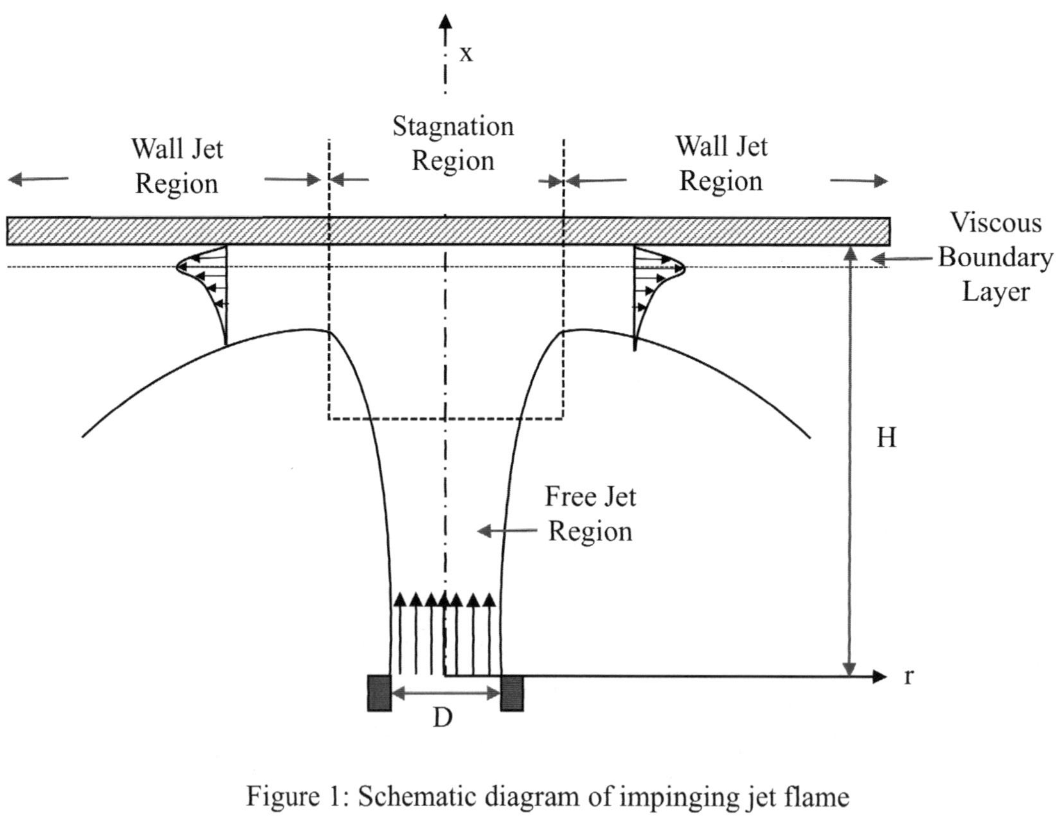


Figure 1: Schematic diagram of impinging jet flame

Journal of Physics: Conference Series **3050** (2025) 012012 doi:10.1088/1742-6596/3050/1/012012

As industries advance toward decarbonisation in alignment with the Paris Agreement, phasing out natural gas in EAFs has become inevitable due to resulting carbon dioxide emissions and their contribution to the furnace's carbon footprint[9, 10]. Hydrogen is emerging as a strong alternative, making its use in EAFs an increasingly viable option[11]. Hence, understanding hydrogen impinging jet flames is crucial to integrating low-carbon alternatives into the EAF.

Literature exists on impinging jet flames [12-15], with most studies focusing on premixed flames - where fuel and oxidiser are mixed before exiting the nozzle. Laminar, transitional, and turbulent flows have been widely investigated through both experimental and numerical methods. Detailed parametric analyses and theories regarding flow patterns in premixed impinging jets can be found in various review papers [12, 15-17]. Despite a deep understanding of premixed impinging jets, non-premixed impinging flames-where fuel and oxidiser are supplied separately and are mixed in the combustion zone, remain under-explored.

In the context of hydrogen impinging jets, while numerous studies have examined the characteristics of premixed hydrogen impinging flames, research on non-premixed hydrogen impinging jets is limited. Ranga-Dinesh et al.[18-21] were among the first to study non-premixed hydrogen impinging jets. The study was conducted through Direct Numerical Simulation (DNS) technique, and the results have shown that near-wall vortical structure plays a crucial role in near-wall heat transfer. Similarly, Martinez et al. [22, 23] implemented the Large Eddy Simulation (LES) approach, focusing primarily on vortex formation near wall boundary regions and heat transfer. The results have shown that the wall jet progresses parallel to the impinging plate, forming large-scale vortices, and are dependent upon the fuel composition. Most of the existing literature emphasises understanding the flow field characteristics of impinging jets, whereas studies on their heat transfer characteristics remain relatively scarce.

He et al.[24, 25] conducted an experiment to analyse the heat transfer characteristics and combustion behaviour of non-premixed impinging flames. The study was focused on $NH_3$ blending in natural gas and understanding its behaviour in reducing emissions such as $NO$, $CO$ and $CO_2$. The experiment was carried out using a non-premixed combustion system combined with an optical measurement system. In the combustion system, fuel ($CH_4$ of 99.9 pct purity) was delivered through a central fuel tube with an 8.0 mm diameter and discharged via a 1.0 mm nozzle. Compressed air (0.4 MPa) was stored in a pressure cylinder and introduced through four symmetrically arranged inlets; the air was conveyed to the combustion chamber via an annular tube with a 40 mm inner diameter, with the nozzle positioned 55 mm from the impinging wall. The optical measurement setup featured a double-pulse Nd-YAG laser (532 nm at 15 Hz), a double-shutter CCD camera (with a resolution of 2360 dpi × 1776 dpi at 16 fps), and a synchronisation controller based on particle image velocimetry (PIV) technology with a timing accuracy of 1 ns. Intercorrelation techniques were applied to process the captured images and calculate velocity vectors, while a high-speed camera recorded the structure of the impinging flame. Figure 2 represents the experimental setup used in this study.

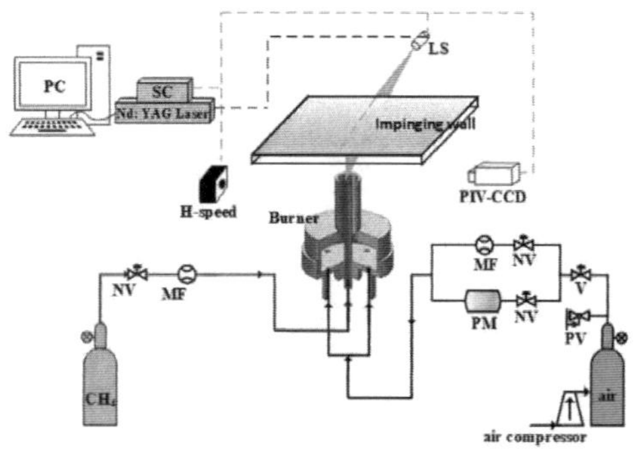


(*a*)

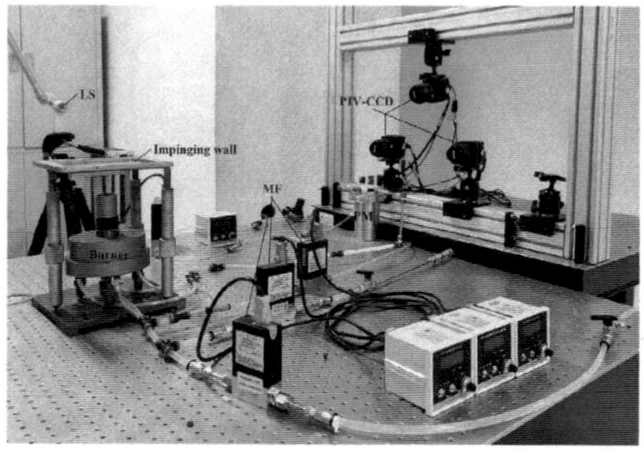


(*b*)

Figure 2: Experimental Setup for optical measurement of an impinging jet flame from He et al. [24, 25]

Many studies suggest that the safe and efficient integration of hydrogen requires significant design and structural modifications to the burner[26]. As a result, blending hydrogen with natural gas is a safer and more practical starting point, as it allows for integration without substantial capital investment. However, there is limited research on hydrogen blending in impinging jet flames. In this study, a mixture of hydrogen and methane is numerically analysed at six different blending ratios. To ensure a constant total heat release across all cases, the fuel flow rate is increased as hydrogen blending dilutes the fuel due to hydrogen's lower density. Air is used as an oxidiser, and the combustion of methane, hydrogen, and air is studied using numerical techniques. The resulting temperature distribution along a steel plate is then computed for each case and compared.

## 2. Methodology

Mathematical modelling principles using computational fluid dynamics (CFD) are followed to investigate the mixture of methane and hydrogen-impinging flames. The following section describes the equations and techniques used to analyse the turbulency and combustion behaviour of the impinging jet.

### i. Governing Equations

There are three primary approaches to modelling turbulent flow: Direct Numerical Simulation (DNS), Large Eddy Simulation (LES), and Reynolds-Averaged Navier-Stokes (RANS) simulation[27]. DNS is the most accurate method, as it resolves turbulence down to the smallest scales, known as the Kolmogorov length scale[28-30]. However, this comes at an extremely high computational cost. On the other hand, the RANS approach is computationally efficient and relatively simple but often introduces significant error margins. To achieve a balance between accuracy and computational feasibility, LES is selected for this study, considering the available computational resources and the required level of precision. In LES, the time-dependent Navier-Stokes (NS) equations—comprising the equations for mass conservation, momentum conservation, energy conservation, and species transport—are filtered using grid spacing, and the resulting equations are then solved. Further details on the Navier-Stokes equations and their theoretical background can be found in the literature [29-31]. In LES, larger eddies are resolved directly and smaller eddies are modelled[27]. A filtered variable can be given as,

$$\bar{\phi}(x) = \int_D \phi(x')G(x, x')dx'$$

Where D is the fluid domain and G is the filter function which determines the scale of the eddies.

The Favre-filtered Navier-Stokes equations can be represented as,

$$\frac{\partial \rho}{\partial t} + \frac{\partial}{\partial x_i}(\rho \bar{u}_i) = 0$$

$$\frac{\partial}{\partial t}(\rho \bar{u}_i) + \frac{\partial}{\partial x_j}(\rho \bar{u}_i \bar{u}_j) = \frac{\partial}{\partial x_j}(\sigma_{ij}) - \frac{\partial \bar{p}}{\partial x_i} - \frac{\partial \tau_{ij}}{\partial x_j}$$

Where, $\sigma_{ij}$ is the stress tensor due to molecular viscosity and can be given as

$$\sigma_{ij} = \left[\mu\left(\frac{\partial \bar{u}_i}{\partial x_j} + \frac{\partial \bar{u}_j}{\partial x_i}\right)\right] - \frac{2}{3}\mu \frac{\partial \bar{u}_j}{\partial x_j}\delta_{ij}$$

And $\tau_{ij}$ is the subgrid-scale stress and can be expressed as,

$$\tau_{ij} = \rho \overline{u_i u_j} - \rho \bar{u}_i \bar{u}_j$$

The subgrid-scale turbulent stresses are modelled through the Boussinesq hypothesis similar to that in the RANS model [32] in which subgrid-scale turbulent stresses were computed from the following relationship

$$\tau_{ij} - \frac{1}{3}\tau_{kk}\delta_{ij} = -2\mu_t \overline{S_{ij}}$$

Where $\mu_t$ is the subgrid-scale turbulent viscosity and is modelled by the dynamic kinetic energy subgrid-scale model proposed by Kim and Menon[33]. The detailed theory related to subgrid-scale modelling can be found in the literature[27, 33, 34].

ii. Physical Problem and Computational Domain

The computational domain is developed to match the experiment conducted by He et al.[24].

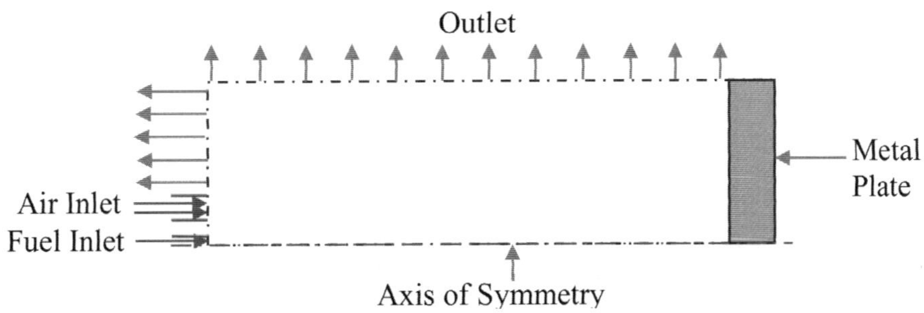


Figure 3: Computational Domain

Journal of Physics: Conference Series **3050** (2025) 012012 doi:10.1088/1742-6596/3050/1/012012

Figure 3 represents the computational domain for the simulation. The geometry consists of a fuel inlet with a diameter of 1 mm, surrounded by annular air inlets with an inner diameter of 20 mm and an outer diameter of 40 mm. The impinging metal plate has a thickness of 4 mm and is positioned 55 mm from the jet. The plate itself has a diameter of 138 mm. The CFD model was developed in a 2D axisymmetric configuration, and Figure 3 depicts the corresponding 2D schematic of the domain. The direction of gravity is opposite to the flame propagation direction.

Three cases are compared in this study. Case I replicates the experimental setup from He et al.[24, 25] for a pure $CH_4$ exactly. Then, in other cases, 10 pct, 25 pct, 50 pct and 75 pct of hydrogen by mass is blended in the fuel, respectively. Adding hydrogen dilutes the fuel; hence, to balance the amount of heat produced, the fuel flow velocity is increased. The total heat content of the fully combusted gas is kept constant in all cases. Hence, each hydrogen blended mixture has a higher fuel inlet velocity.

Table 1: Boundary parameters used in the computation

| | | **Boundary Type** | **Values and Composition** |
|---|---|---|---|
| Fuel Inlet | Case I | Velocity | $8\ m/s$ , $100\%\ CH_4$ |
| | Case II | Velocity | $11.91\ m/s$, $90\% CH_4 + 10\%\ H_2$ |
| | Case III | Velocity | $16.25\ m/s$, $75\%\ CH_4 + 25\%\ H_2$ |
| | Case IV | Velocity | $21.10\ m/s$, $50\%\ CH_4 + 50\%\ H_2$ |
| | Case V | Velocity | $24.29\ m/s$, $25\% CH_4 + 75\%\ H_2$ |
| | Case VI | Velocity | $26.56\ m/s$, $100\%\ H_2$ |
| | Case VII | Velocity | $8\ m/s$, $100\%\ H_2$ |
| Oxidiser Inlet | Case I | Velocity | $1.1\ m/s$ |
| | Case II | Velocity | $1.06\ m/s$ |
| | Case III | Velocity | $1.02\ m/s$ |
| | Case IV | Velocity | $0.97\ m/s$ |
| | Case V | Velocity | $0.94\ m/s$ |
| | Case VI | Velocity | $0.91\ m/s$ |
| | Case VIII | Velocity | $0.28\ m/s$ |
| Outlet | | Pressure outlet | 101.325 KPa |
| Wall boundary (metal plate) | | Mixed boundary | $h_{ext} = 10\ W/m^2K$<br>$\varepsilon_{ext} = 0.94$,<br>$T_{ext} = 300K$,<br>$T_{\infty} = 300K$ |

Mixture fraction relationship:

A simple one-step reaction mechanism is taken as a reference, and based on the lower heating value (LHV) of methane and hydrogen, mixture velocity and density are calculated for different hydrogen blended ratios.

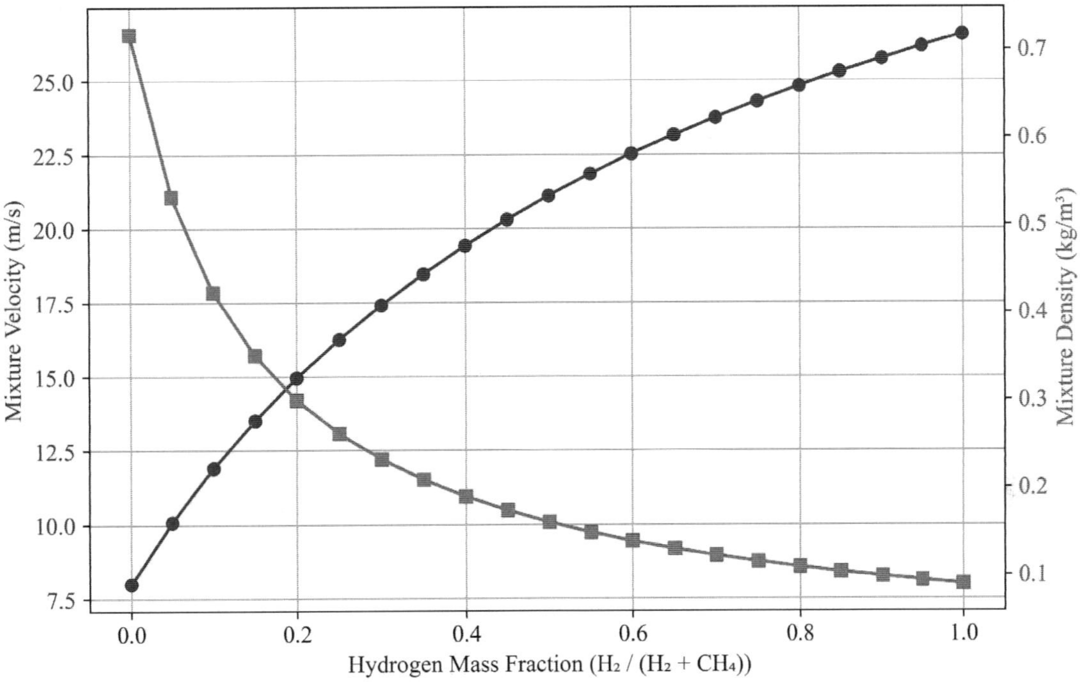


Figure 4: Hydrogen-Methane mixture relationship while keeping constant energy

Since hydrogen and methane have significantly different densities and lower heating values (LHV), adding hydrogen to methane by mass does not produce a linear change in the mixture's properties. As the hydrogen mass fraction increases, the mixture density drops rapidly, so the velocity needed to deliver a constant heat release rate rises sharply. In an ideal gas mixture, the density is directly proportional to the average molar mass of the mixture. When hydrogen is added by mass, due to its significantly lower molar mass compared to methane, the number of moles per unit mass increases. This alters the average molar mass non-linearly, which in turn affects the mixture density. As described by the ideal gas law, this results in a non-linear relationship between hydrogen mass fraction and the mixture density. The power for each case was kept at 225.5 Watts. This behaviour is illustrated in Figure 4.

iii. Computational Procedure, Meshing and Convergence

Time-dependent filtered fluid flow governing equations, as explained in section i, were solved using ANSYS fluent 22R1 software. Initially, RANS equations were solved in steady state mode using the $k - \omega$ Shear Stress Transport (SST) turbulence model. The steady-state results were then used to initialise the Large-Eddy Simulation (LES) model. A transient flow simulation is conducted with a timestep size of 0.005 seconds and for 50 timesteps. Then, unsteady mean variables are collected through sampling of instantaneous results. A dynamic kinetic-energy model with the viscous heating option was employed for subgrid-scale modelling. For radiation modelling, the discrete ordinates (DO) radiation model was utilised[27]. To capture the complex combustion phenomenon, a non-premixed steady diffusion flamelet model was adopted. The GRIMech3.0 combustion mechanism[35], which consists of 50 species, along with its associated thermodynamic property file, was imported to generate a probability density function (PDF) table. The theoretical background on the non-adiabatic steady diffusion model can be found in the ANSYS Fluent Theory Guide[27].

To determine the appropriate mesh size for the wall boundary layer, a refined mesh was generated in the near-wall region using sizing controls. A mesh independence study was conducted by evaluating three different cases: coarse, fine, and extra-fine meshes with mesh sizes of 9328, 117,680, and 196,470 elements respectively. Flow parameters were plotted for each case, and the errors between the fine and extra-fine meshes were found to be well below 1 pct. Considering the computational cost, the fine mesh was selected for the rest of the simulations. Convergence was achieved when all residuals dropped below 0.0001 and the mass flow rate at the outlet remained steady. Figure 5 illustrates the transient average velocity profiles for grid sizes categorised as coarse, fine, and extra fine.

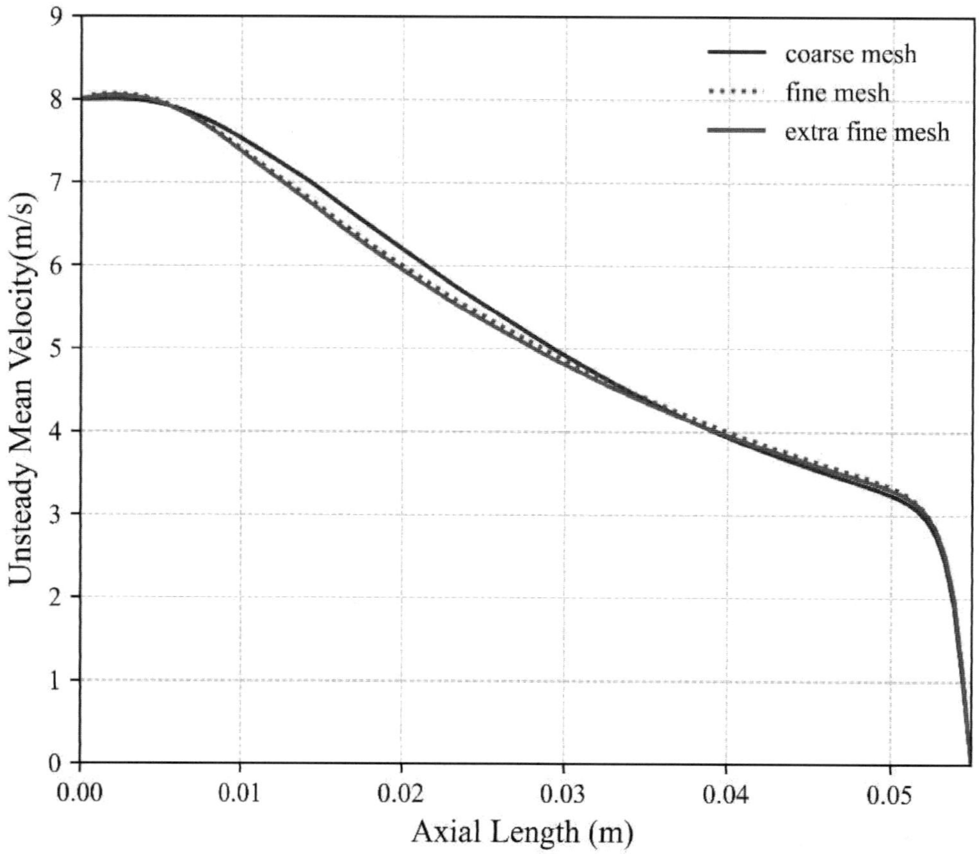


Figure 5: Unsteady mean velocity at different mesh sizes

## 3. Results and Discussions

### i. Flow field and Temperature validation

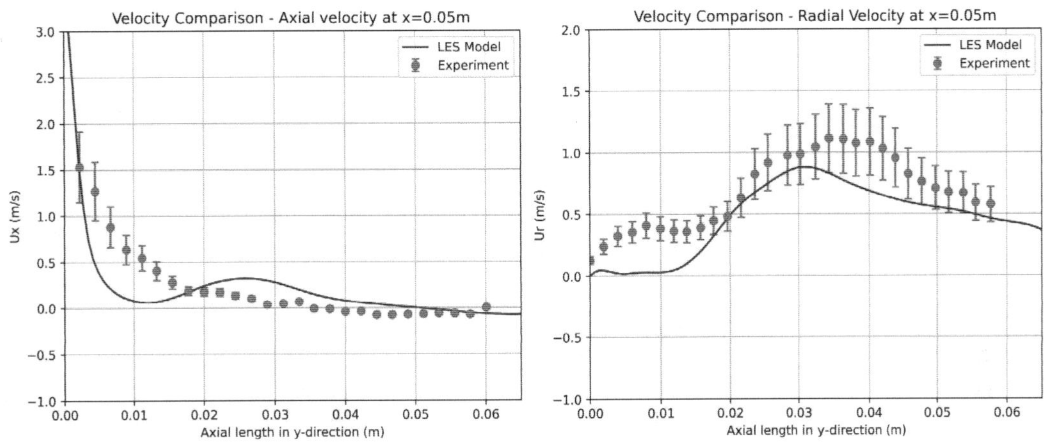


Figure 6: Mean velocity magnitude at x = 0.05m (a) axial velocity (b) radial velocity

The LES-based CFD results were validated against the experimental results for impinging flames with methane fuel. The axial and radial velocities in the y-direction at x = 0.05m were sampled at each time step, and the unsteady mean velocities were calculated. As shown in Figure 6, the axial velocity shows a good agreement with the experimental results[24, 25, 36], while the radial velocities display a promising pattern. The discrepancies observed may stem from experimental error margins, which are also noted in the experiment. Moreover, since the velocity measurements were taken in the highly turbulent region, the LES approach, which directly resolves large-scale eddies but models the smaller, fine-scale motions, may not have fully captured the complex flow structures at the smallest scales.

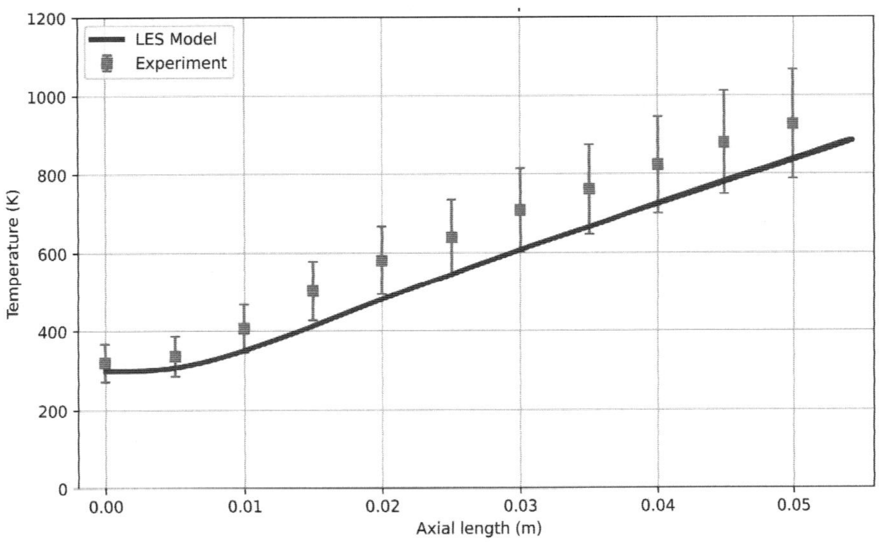


Figure 7: Axial temperature profile[24, 25]

Journal of Physics: Conference Series **3050** (2025) 012012 doi:10.1088/1742-6596/3050/1/012012

The flame temperature along the central axis of the flame was also compared with the experimental results as shown in Figure 7. The modelled temperature profile along the central axis of the flame shows a good agreement with the experimental results with the provided experimental error margin.

ii. Flow field characteristics

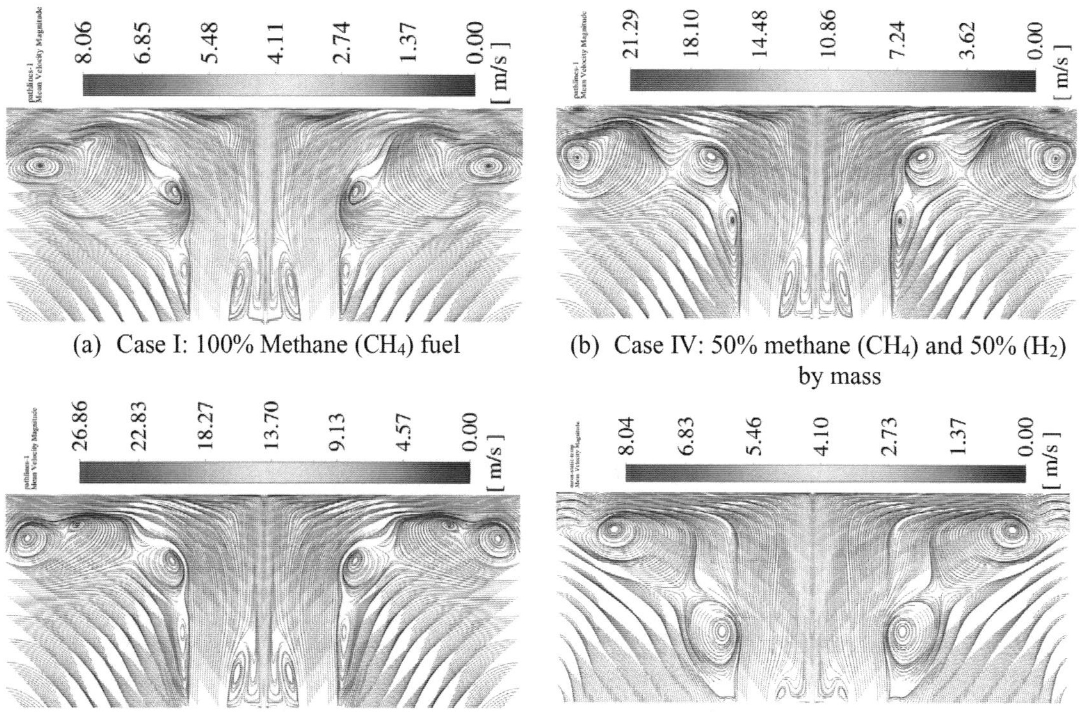


(a) Case I: 100% Methane ($CH_4$) fuel

(b) Case IV: 50% methane ($CH_4$) and 50% ($H_2$) by mass

(c) Case VI: 100% Hydrogen ($H_2$) fuel

(d) Case VII: Hydrogen ($H_2$) with same flowrate as Case I

Figure 8: Velocity path-lines for Case (a) I, (b) IV, (c) VI, and (d) VII

Figure 8 illustrates the fluid particle pathlines coloured with unsteady mean velocity magnitude. The CFD results predict that the addition of hydrogen in the fuel tends to form a higher number of vortical structures towards the impingement surfaces. However, as shown in Figure 8 (d), fewer vortical structures are likely to form which is mainly due to the hydrogen's lower density and combustion occurring closer to the nozzle at the lower velocity. The vortical structures primarily smaller eddies formed due to the addition of hydrogen which can be observed in Figure 8 (b) and (c). These smaller eddies are known to promote the heat transfer from the flame to the metal plate[22, 23].

Journal of Physics: Conference Series **3050** (2025) 012012
doi:10.1088/1742-6596/3050/1/012012

### iii. Temperature Distribution

a. Temperature distribution in the flow field

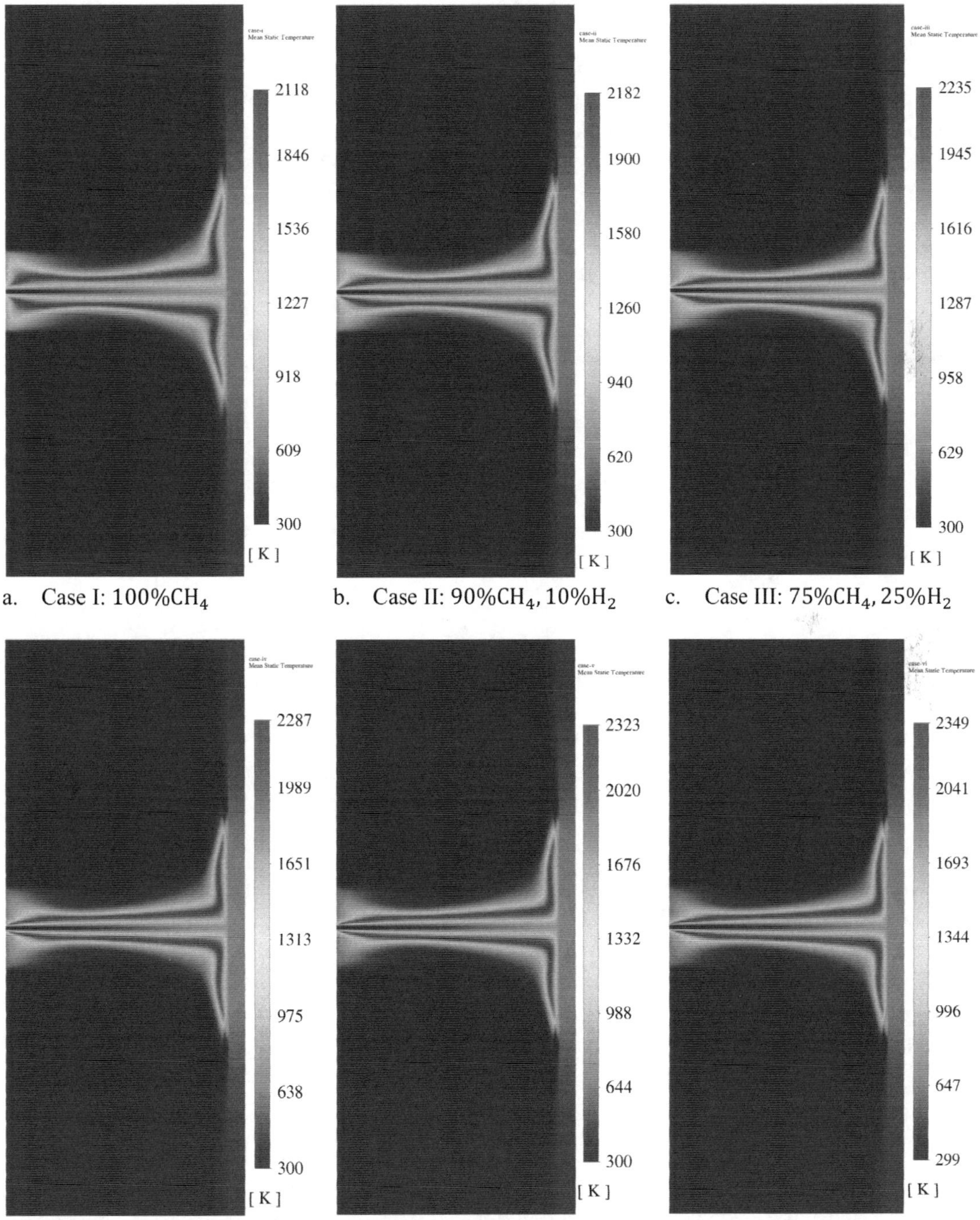


a. Case I: $100\%CH_4$ b. Case II: $90\%CH_4, 10\%H_2$ c. Case III: $75\%CH_4, 25\%H_2$

d. Case IV: $50\%CH_4, 50\%H_2$ e. Case V: $25\%CH_4, 75\%H_2$ f. Case VI: $100\%H_2$

Figure 9: Temperature contours for different cases

Figure 9 shows the temperature contours with the blending of hydrogen in the methane fuel. The temperature contours in each case are influenced by the flow characteristics and fuel properties. The peak temperature in hydrogen-impinging flames gradually increased from 2118K to 2349K, an increase of 231K, as the hydrogen content ranged from 0 pct in Case I to 100 pct in Case VI. This increase in temperature in metal plates may be attributed due to higher flame temperature and better heat transfer through convection in hydrogen-rich flames. Furthermore, it is predicted that a low radiative heat loss from the hydrogen-rich flames in the free jet region could explain why the metal plate to a higher temperature than that of methane-rich flames. As the flow rate was increased to maintain the steady heat release from the flame in each case, the hot region in the wall-jet region remained approximately unchanged, despite the fuel mixture being less dense. The stagnation region and wall-jet interaction region are crucial aspects of impinging jets, as the majority of heat transfer occurs through convection in this area, ensuring efficient thermal energy transfer.

b. Temperature distribution in the steel plate

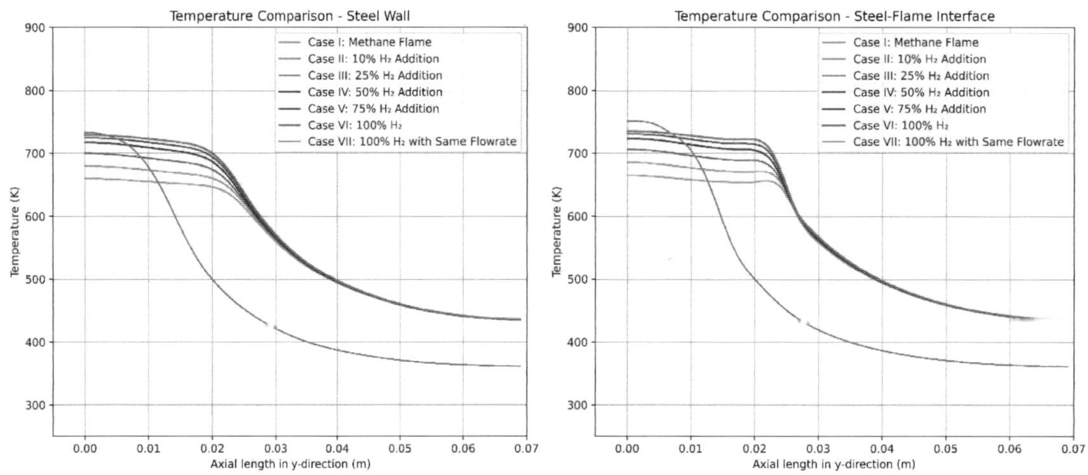


Figure 10: Temperature profile along the steel plate at (a) outer wall and (b) steel-flame interface

The temperature profiles along the inner and outer surfaces of the impinging plate were plotted as shown in Figure 10. Since hydrogen flames burn at higher temperatures when combusting with air, the temperature at the steel plate is also noticeably higher. Despite hydrogen being supplied at a much higher velocity than methane, the flame impingement region remains similar to that of methane, as seen in Figure 10 (right). The temperature at the impinging steel plate can reach up to 735 K for the hydrogen flame (Case VI) compared to 665 K for the methane flame (Case I). The CFD results predict that as the hydrogen content increases in the fuel mixture, the temperature along the plate gradually increases.

Although hydrogen flames have lower radiative characteristics, convection plays a dominant role in distributing heat along the steel plate. However, in the case of a hydrogen-impinging flame supplied at the same velocity as methane (Case VII), the stagnation region is significantly smaller, resulting in a heating region approximately half the size of the methane flame impingement region. In Case VII, the lower radiative properties of hydrogen become more evident, as the temperature along the plate drops at a much faster rate compared to other cases.

iv. Species mass fractions

c. $CO_2$ mass fractions

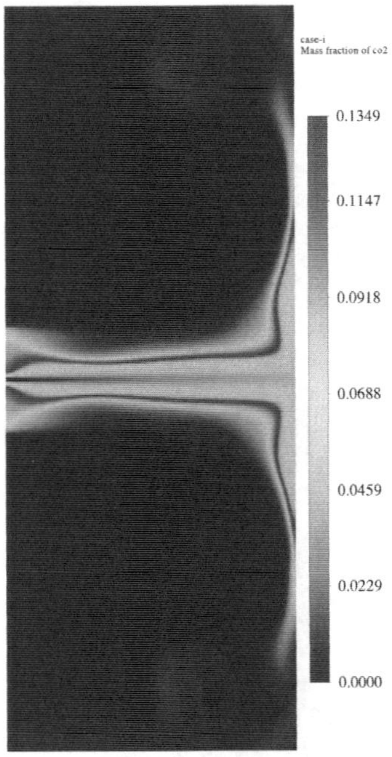


a. Case I: $100\%CH_4$

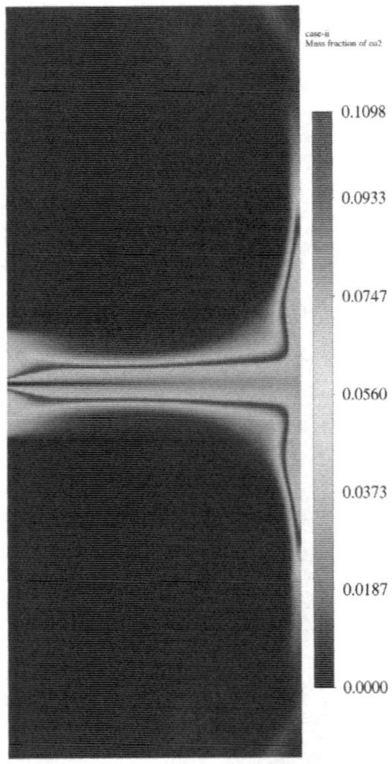


b. Case II: $90\%CH_4, 10\%H_2$

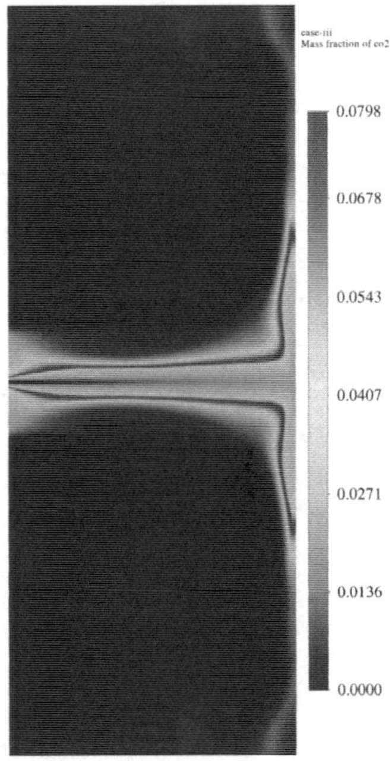


c. Case III: $75\%CH_4, 25\%H_2$

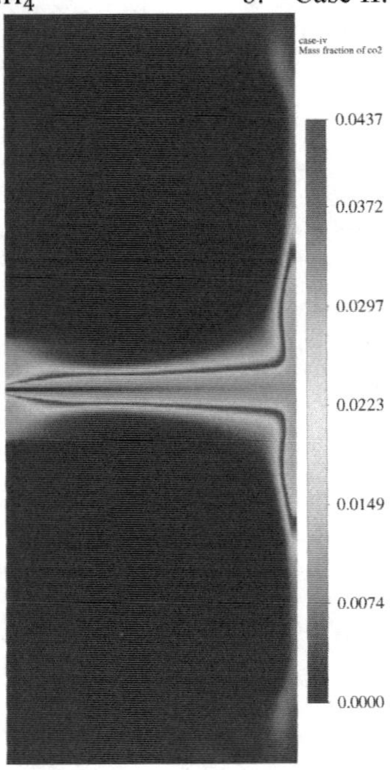


d. Case IV: $50\%CH_4, 50\%H_2$

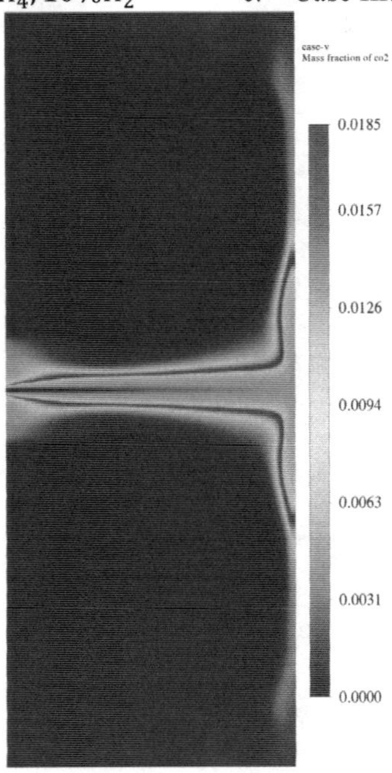


e. Case V: $25\%CH_4, 75\%H_2$

Figure 11: $CO_2$ mass fractions for different cases

Journal of Physics: Conference Series **3050** (2025) 012012 doi:10.1088/1742-6596/3050/1/012012

Figure 11 shows the prediction of carbon dioxide ($CO_2$) in different hydrogen blended cases. The CFD results predict that the $CO_2$ mass fractions will gradually decrease as the hydrogen content increases in the fuel mixture. The concentration of $CO_2$ distribution along the flame region has also gradually reduced as the hydrogen content increased, however, the shape of the distribution remains unchanged.

d. $H_2O$ mass fractions

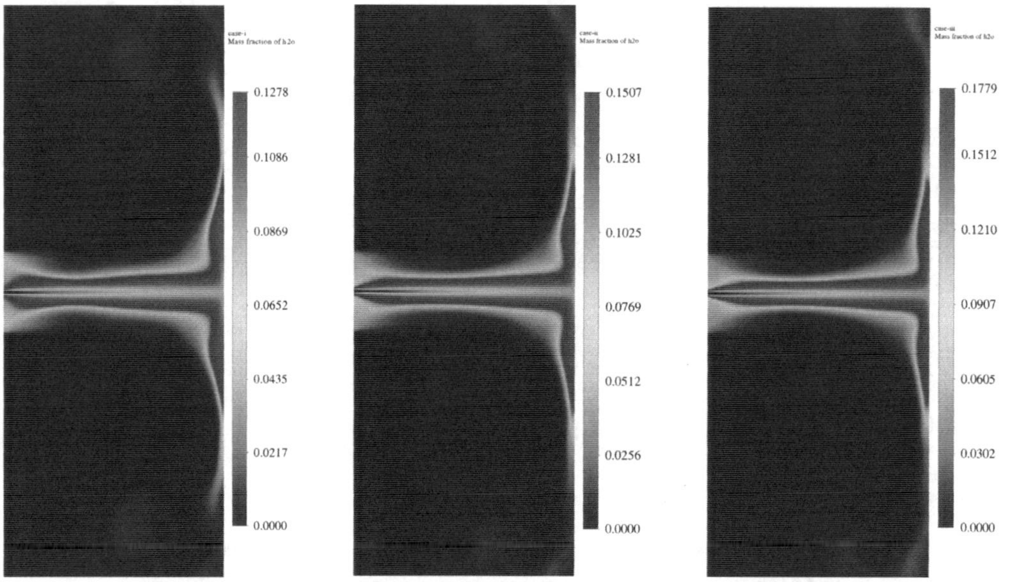


a. Case I: $100\%CH_4$ b. Case II: $90\%CH_4, 10\%H_2$ c. Case III: $75\%CH_4, 25\%H_2$

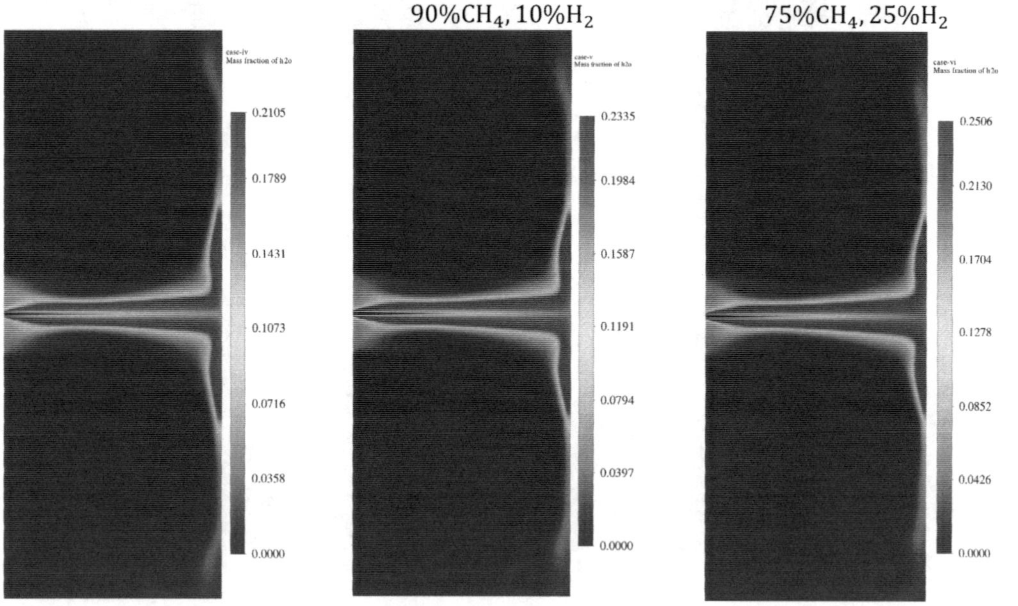


d. Case IV: $50\%CH_4, 50\%H_2$ e. Case V: $25\%CH_4, 75\%H_2$ f. Case VI: $100\%H_2$

Figure 12: $H_2O$ mass fractions for different cases

The major combustion product of natural gas fuel is carbon dioxide whereas the product of hydrogen fuel is only water vapour. Figure 12 shows the contours of $H_2O$ distribution during the combustion. The CFD model predicts that the addition of hydrogen in the fuel mixture increases the formation of $H_2O$ in the combustion zone, however the area of $H_2O$ distribution remains comparatively unchanged.

## 4. Conclusion and Recommendations

Large Eddy Simulation (LES) is performed to analyse the effect of hydrogen blending in natural gas in flame impingement conditions. In metal heating operations, especially in steelmaking furnaces such as EAFs and Reheat Furnaces, flame impingement is a dominant means of heat transfer from flames to metals. Since non-premixed burners are predominant in the steel industry, this study explores the flow field and heat transfer characteristics of non-premixed impinging flames using methane, hydrogen and a mixture of both as fuels at different ratios. Based on this study, the following conclusions can be drawn.

- Hydrogen blending enhances the flame temperature gradually. The temperature attained by the hydrogen flame was approximately 2349K, in contrast to 2118K for methane-impinging flames, assuming both flames generate equivalent heat energy.
- Increased hydrogen blending in the natural gas reduces carbon dioxide ($CO_2$) produced and increases the amount of water vapour ($H_2O$) produced.
- The stagnation region on the impingement surface attains higher temperatures of 735K for hydrogen-impinging flames, in contrast to 665K for methane-impinging flames. In different blending cases, as the hydrogen content increases, the surface temperature increases gradually.
- Despite fuel being supplied at a higher velocity in all the blending cases as compared to the velocity of methane to maintain the constant heat release from the flame, the primary flame impingement region remains relatively unaffected in all cases.
- When hydrogen is supplied at the same velocity as methane, the resulting flame is significantly smaller, with an impingement region approximately half the size of that produced by methane.

The LES-based CFD model is validated only against the impinging flame with methane fuel and the hydrogen blending cases are further explored with CFD models. Numerical simulations by solving the Navier–Stokes equations require capturing a wide range of time and length scales that influence the flow field. One way to capture these details is through direct numerical simulation (DNS), however, it is computationally demanding. Therefore, the LES technique is employed in this study, relying on scale separation. LES resolves the large scales that carry most of the energy and govern flow transport, while the smaller, more universal scales are modelled. To further explore the broad range of time and length scales, DNS can be implemented, or experimental studies can be conducted to build a deeper understanding of the heat transfer, heating, and radiative characteristics of hydrogen-impinging flames.

## References


1. Worrell, E., *Advanced technologies and energy efficiency in the iron and steel industry in China.* Energy for Sustainable Development, 1995. **2**(4): p. 27-40.
2. Mandal, K. and G.A. Irons, *A study of scrap heating by burners. Part I: experiments.* Metallurgical and Materials Transactions B, 2013. **44**: p. 184-195.
3. Thomson, M., et al., *Effect of oxyfuel burner ratio changes on energy efficiency in electric arc furnace at Co-Steel Lasco.* Ironmaking & steelmaking, 2001. **28**(3): p. 266-272.
4. De Beer, J., E. Worrell, and K. Blok, *Future technologies for energy-efficient iron and steel making.* Annual review of energy and the environment, 1998. **23**(1): p. 123-205.
5. Worrell, E., *Energy efficiency improvement and cost saving opportunities for the US iron and steel industry an ENERGY STAR (R) guide for energy and plant managers.* 2011.
6. Mandal, K., *Modeling of scrap heating by burners.* 2010.
7. Pandey, G., et al., *NOx Prediction of Supersonic Coherent Jets for Electric Arc Steelmaking Furnace.* Metallurgical and Materials Transactions B, 2024.
8. Mathur, P., *Coherent jets in steelmaking: principles and learnings.* Praxair Met Technol, 2004. **47**(3): p. 184-205.
9. Bataille, C., et al., *A review of technology and policy deep decarbonization pathway options for making energy-intensive industry production consistent with the Paris Agreement.* Journal of Cleaner Production, 2018. **187**: p. 960-973.
10. Hornby, S. and G. Brooks, *Impact of Hydrogen DRI on EAF Steelmaking: MIDREX Report-Second Quarter 2021 Available at: https://www. midrex. com/tech-article/impact-of-hydrogen-dri-on-eaf-steelmaking.* Accessed on 27/10, 2021.
11. Mc, *Hydrogen burner | E&M Combustion | Industrial burners.* 2019.
12. Chander, S. and A. Ray, *Flame impingement heat transfer: a review.* Energy conversion and Management, 2005. **46**(18-19): p. 2803-2837.
13. Baukal Jr, C. and B. Gebhart, *A review of empirical flame impingement heat transfer correlations.* International Journal of Heat and Fluid Flow, 1996. **17**(4): p. 386-396.
14. Gauntner, J.W., *Survey of literature on flow characteristics of a single turbulent jet impinging on a flat plate.* 1970.
15. Viskanta, R., *Heat transfer to impinging isothermal gas and flame jets.* Experimental Thermal and Fluid Science, 1993. **6**(2): p. 111-134.
16. Polat, S., et al., *Numerical flow and heat transfer under impinging jets: a review.* Annual review of heat transfer, 1989. **2**.
17. Baukal, C.E. and B. Gebhart, *A review of semi-analytical solutions for flame impingement heat transfer.* International Journal of Heat and Mass Transfer, 1996. **39**(14): p. 2989-3002.
18. Ranga Dinesh, K.K.J., X. Jiang, and J.A. van Oijen, *Numerical simulation of hydrogen impinging jet flame using flamelet generated manifold reduction.* International Journal of Hydrogen Energy, 2012. **37**(5): p. 4502-4515.
19. Ranga Dinesh, K.K.J., X. Jiang, and J.A. van Oijen, *Hydrogen-enriched non-premixed jet flames: Analysis of the flame surface, flame normal, flame index and Wobbe index.* International Journal of Hydrogen Energy, 2014. **39**(12): p. 6753-6763.

20. Ranga Dinesh, K.K.J., X. Jiang, and J.A. van Oijen, *Hydrogen-enriched non-premixed jet flames: Compositional structures with near-wall effects.* International Journal of Hydrogen Energy, 2013. **38**(12): p. 5150-5164.
21. Ranga Dinesh, K.K.J., et al., *Hydrogen-enriched nonpremixed jet flames: Effects of preferential diffusion.* International Journal of Hydrogen Energy, 2013. **38**(11): p. 4848-4863.
22. Mira Martinez, D. and X. Jiang, *Numerical investigations of a hydrogen impinging flame with different finite-rate chemical kinetic mechanisms.* Fuel, 2013. **109**: p. 285-296.
23. Mira Martinez, D., et al., *Numerical investigation of the effects of fuel variability on the dynamics of syngas impinging jet flames.* Fuel, 2013. **103**: p. 646-662.
24. He, C., et al., *Analysis of the NH3 blended ratio on the impinging flame structure in non-premixed CH4/NH3/air combustion.* Fuel, 2022. **330**: p. 125559.
25. He, C., et al., *Analysis of the quenching behavior in impinging flame: Flow and thermal characteristics.* Numerical Heat Transfer, Part A: Applications, 2024. **85**(12): p. 1973-1987.
26. Palacios, A. and D. Bradley, *Conversion of natural gas jet flame burners to hydrogen.* International Journal of Hydrogen Energy, 2021. **46**(33): p. 17051-17059.
27. Fluent, A., *Ansys fluent theory guide.* Ansys Inc., USA, 2011. **15317**: p. 724-746.
28. Alam, M., et al., *Computational fluid dynamics modeling of supersonic coherent jets for electric arc furnace steelmaking process.* Metallurgical and Materials Transactions B, 2010. **41**: p. 1354-1367.
29. Versteeg, H.K., *An introduction to computational fluid dynamics the finite volume method, 2/E.* 2007: Pearson Education India.
30. Batchelor, G.K., *An introduction to fluid dynamics.* 2000: Cambridge university press.
31. Pandey, G., et al., *NOx Prediction of Supersonic Coherent Jets for Electric Arc Steelmaking Furnace.* Metallurgical and Materials Transactions B, 2024. **55**(5): p. 3332-3345.
32. Chen, C.J., *Fundamentals of turbulence modelling.* 1997: Crc Press.
33. Kim, W.-W., et al. *Application of the localized dynamic subgrid-scale model to turbulent wall-bounded flows.* in *35th aerospace sciences meeting and exhibit.* 1997.
34. Kim, W.-W. and S. Menon. *A new dynamic one-equation subgrid-scale model for large eddy simulations.* in *33rd aerospace sciences meeting and exhibit.* 1995.
35. Gregory P. Smith, D.M.G.M.F.N.W.M.B.E.M.G.C.T.B.R.K.H.S.S.W.C.G.J.V. and Q. Zhiwei. *GRI-Mech 3.0.* 2023; Available from: http://combustion.berkeley.edu/gri-mech/version30/text30.html.
36. Nikolaou, Z.M., J.-Y. Chen, and N. Swaminathan, *A 5-step reduced mechanism for combustion of CO/H2/H2O/CH4/CO2 mixtures with low hydrogen/methane and high H2O content.* Combustion and flame, 2013. **160**(1): p. 56-75.

# AUTHOR INDEX